Zhiqun Cheng, Guohua Liu
Communication Electronic Circuits

Also of interest

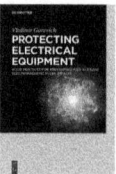

Zhiqun Cheng, Guohua Liu

Communication Electronic Circuits

—

DE GRUYTER   Science Press
Beijing

Main Authors
Zhiqun Cheng
School of Electronic Information
Hangzhou Dianzi University
zhiqun@hdu.edu.cn

Guohua Liu
School of Electronic Information
Hangzhou Dianzi University

Co-Authors
Huajie Ke
School of Electronic Information
Hangzhou Dianzi University

Mi Lin
1158# No.2 St., Xiasha Universty Park,
Hangzhou, China. 310018

Jin Chen
1158# No.2 St., Xiasha Universty Park,
Hangzhou, China. 310018

Zhihua Dong
1158# No.2 St., Xiasha Universty Park,
Hangzhou, China. 310018

Tao Zhou
1158# No.2 St., Xiasha Universty Park,
Hangzhou, China. 310018

ISBN 978-3-11-059538-3
e-ISBN (PDF) 978-3-11-059382-2
e-ISBN (EPUB) 978-3-11-059293-1

Library of Congress Control Number: 2020935440

Bibliographic information published by the Deutsche Nationalbibliothek
The Deutsche Nationalbibliothek lists this publication in the Deutsche Nationalbibliografie;
detailed bibliographic data are available on the Internet at http://dnb.dnb.de.

Introduction

This textbook comprises six chapters: introduction; radiofrequency amplifiers; sinusoidal oscillators; amplitude modulation, demodulation and frequency mixing; angle modulation and demodulation; and phase lock loop. Each chapter is narrated in the logical sequence of concept introduction, circuit composition, mechanism, detailed calculation or analysis, and circuit applications. To encourage better understanding, well-designed examples and problems are included in each chapter for readers.

We looked through various domestic and international textbooks of related subjects during years of teaching experience. While extracting the knowledge essence and emphasizing in-depth and comprehensive understanding of concepts and principles with utmost significance, we feel urgent to update mainstream applications and help our readers keep up with the modernization of technology in the mean time. This book can serve as a preliminary introduction to the world of communication electronic circuits for engineering majors in university education. It also works as a good handbook for researchers and engineers.

https://doi.org/10.1515/9783110593822-202

Preface

A communication electronic circuit is one of the core courses of electronic information major. This course aims at helping students master the composition, principle, properties and analysis methodology of basic functional circuits. With the independent parallel experiment course, students are also trained for the techniques of designing, installation and testing those circuits along with the ability of circuit analysis, design and application. It helps students build solid and necessary foundation for future and practical work. The book covers the transitional area between low-frequency and high-frequency wireless circuits. Specifically, it introduces the fundamental physical principles related to the operation of a typical wireless radio communication system and functional circuit modules. It is different from other textbooks that we are not attempted to discuss all the possible topics related to wireless communication circuits and system. However, many textbooks cover wireless communication systems comprehensively but with relatively few details, which are not suitable for students' in-depth self-study. In this textbook, we have chosen some topics to discuss in more depth, and thus provide detailed mathematical derivations, applied approximations and analogies. In our experiences, the chosen topics are suitable for a one semester, 3 h per week, senior undergraduate engineering major. Our intention is to provide a logical relation that flows smoothly from one chapter to the next, hoping that the readers will find it easy to understand it. Our main inspiration in writing this English textbook came from the requirement of bilingual teaching for the students of our school, in Hangzhou Dianzi University. Some authors write textbook covering many topics at a high level. Other similar books cover fewer fundamental principles with more details. We choose the latter according to our undergraduate level. All of the contents in this textbook are focused on the basic concepts and fundamental radio-frequency circuits working principles, which will provide necessary knowledge and abilities for engineering students entering the field of wireless communication electronics. Therefore, the intended potential readers for this book are, primarily, senior undergraduate electronic engineering students preparing for their careers in communication electronics. Meanwhile, our hope is that graduate engineering students will find this book a useful reference.

The book covers the following topics:

(1) Introduction to the basic knowledge of the communication electronics circuits. This chapter serves as a preliminary introduction to the communication electronic circuit system. Contents include the working principle of basic receiver and transmitter units, operating principle of LC resonant circuit including resonance condition, resonant characteristics, resonant curves and the functions in the circuit, as well as characteristics of nonlinear devices.

(2) Radiofrequency amplifiers. Both small signal tuned amplifiers and class-C large power amplifiers are analyzed in detail. High-efficiency power amplifier class-E, class-F and class-EF power amplifiers are briefly introduced as well.

https://doi.org/10.1515/9783110593822-203

(3) Sinusoidal oscillators. In this chapter, we study various types of oscillators. The main focus is the understanding of principles of different oscillators, together with the comparison of their advantages and disadvantages.

(4) Amplitude modulation, demodulation and frequency mixing. The mechanism and typical circuits of amplitude modulation and demodulation processes are discussed.

(5) Angle modulation and demodulation. The mechanism and typical circuits of frequency/phase modulation and demodulation processes are discussed.

(6) Phase lock loop. It discusses the important control circuits in modern communication electronics of phase lock loop including its basic principle, circuit composition, mathematical model and applications.

This textbook is originated by our group of "Communication Electronics Circuits Course" of Hangzhou Dianzi University at Hangzhou, China.

The authors of this textbook are Zhiqun Cheng, Guohua Liu, Mi Lin, Jin Chen, Zhihua Dong, Huajie Ke and Tao Zhou. They are all from Hangzhou Dianzi University.

Zhiqun Cheng et al.

Contents

Chapter 1
Introduction to basic knowledge of communication electronic circuits

1.1 Introduction to communication system

Communication is a process of delivering information from transmitter to receiver. All the required technical equipments and transmission medium during the process compose a communication system altogether. There are various approaches to realize information communication. The history can be traced back to the ancient times when the primitive methods were used widely such as beacons, courier stations and signal fires. The study on electromagnetism in the seventeenth century laid the theoretical foundation for modern communication. Samuel Morse invented the first telegraph device in 1837, which practically marked the beginning of modern electrical communication. Being fast, accurate, reliable and massive in global coverage, modern communication systems are playing irreplaceable roles in all aspects of life.

1.1.1 Composition of communication system

Modern communication systems are mainly in the forms of voltage signals, current signals, electromagnetic (EM) waves or light signals. In this book, communication refers to electrical communication. The general model of communication system is shown in Fig. 1.1.

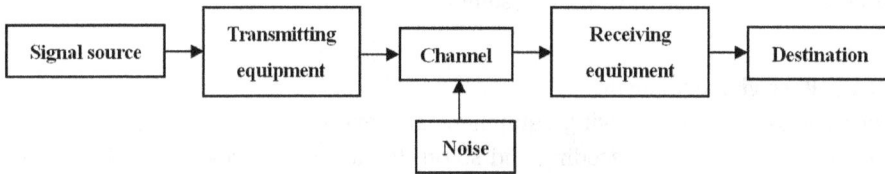

Fig. 1.1: General model of communication system.

1.1.1.1 Signal source
Signal source is the origin of information, which can be analog or digital. Analog source outputs continuous signals, such as telephone and television. Digital source outputs discrete signals, such as computer and other digital terminal equipments.

The output signal of signal source is called baseband signal or modulation signal. Baseband signal (or modulation signal) refers to the original signal with the

https://doi.org/10.1515/9783110593822-001

frequency spectrum starting from zero to low frequencies. Accordingly, baseband signal can be analog or digital.

1.1.1.2 Transmitting equipment

Transmitting equipment functions to match signal source and channel. By transforming and processing baseband signals, information becomes possible for transmission in channel. The original messages are not always electrical, such as sound, text and image, and they should be converted into electrical signals by the transmitting equipment. For instance, microphone converts voice signals into continuous analog electrical signals and then transmitting equipment sends the analog electrical signals to the channel. For digital signals, transmission equipments often contain source coding and channel coding units.

1.1.1.3 Channel

Channel is the path for information transmission like a highway for automobiles. And it usually includes unwanted interferences or noises at the same time. The property of transmission material and interference are closely related to the quality of communication. Channel can be wired or wireless, both of which have many choices on transmission media. Wired channels include electrical cable, optical cable and so on. Wireless channels include the surface of earth – ground, seawater and air. In space communication, channel is the free space of the universe.

1.1.1.4 Noise

Noise refers to all unwanted interferences and distortions that occur in channel or other communication units of the system.

1.1.1.5 Receiving equipment

The functions of the receiving equipment are reverse to transmitting equipment, namely, demodulation, decoding and so on. Its task is to recover the original signals based on the received signals with interferences. For example, telephones can restore transmitted electrical signals back into human sounds. Since interferences and distortions exist in every process of signal transmission and recovery, receiving equipments are expected to minimize those negative effects.

1.1.1.6 Destination

Destination is the final recipient of information.

1.1.2 Classification of communication systems

Communication systems are categorized differently based on different standards. Although the system composition and equipments are not quite the same, the operational principle is usually similar.

(1) According to transmission channel, communication systems can be divided into wired and wireless. Wired communication utilizes transmission channels such as electric cable, optical cable and waveguide which are physical media. In contrary, wireless communication utilizes air or free space to transmit signals. Common forms of wireless communication are microwave communication, shortwave communication, mobile communication, satellite communication, scatter communication and laser communication.

(2) According to the types of signals, communication systems can be divided into analog or digital systems.

(3) According to modulation, communication systems can be different as baseband transmission or waveband (modulation) transmission. Baseband transmission transmits signals directly without modulation, such as local audio telephone; waveband transmission deals with modulated signals.

(4) According to signal transmission mode, communication systems can be divided into three categories, simplex, half duplex and duplex. Simplex is a one-way transmission mode in which messages are delivered with no response needed. Radio, remote controller and television are familiar examples of simplex communication. Half duplex communication means messages can be sent and received from both sides, but the receiving and the sending action cannot take place at the same time. Intercoms and transceivers are typical half duplex communication. Duplex communication is a two-way transmission mode. In this mode, messages can be sent and received simultaneously such as telephone and mobile phone.

(5) According to terminal movement, communication systems can be divided into mobile or fixed. Mobile communication is that at least one terminal is in motion.

(6) According to physical characteristics of information, there are telephone, telegraph, fax communication systems, radio and television communications systems, data communication systems and so on.

(7) According to operating frequency, communication systems can be divided into longwave, medium wave, shortwave and microwave communication systems.

(8) Additionally, there are also other classifications. For example, according to multiple addresses mode, there are frequency division multiple access communication, time division multiple access communication and code division multiple access communication. According to users, there are public communication and private communication. According to location of the communication objects, there are ground communication, air communication, deep space communication, underwater communication and so on.

In this book, we mainly focus on wireless analog communication system and explain the circuit composition, operational principle, theoretical and practical analysis and application of each important communication units.

1.2 Basic components of communication system

A wireless communication circuit includes three basic parts, transmitting unit, channel and receiving equipment.

1.2.1 Transmitting unit

Transmitting unit must include a transducer, a transmitter and an antenna. Its diagram is shown in Fig. 1.2.

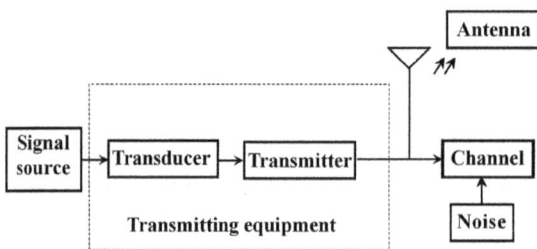

Fig. 1.2: Diagram of a transmitting unit.

The core part of a transmitting unit is transmitter. Transmitter realizes modulation and power amplification of signals. Main performance indices of transmitter are spectrum width of the modulation signals, operating frequency, output power, efficiency, frequency stability, spurious emission, out-of-band noise, in-band spurious and so on.

Transducer converts the raw nonelectrical signals into electric signals. The electric signals are typically at low frequencies and known as modulation or baseband signals. However, the low-frequency electric signals cannot be transmitted directly via antenna. Only when the wavelength is close to the size of antenna, signal can be radiated effectively and then propagate in the form of EM wave. As an example, the frequency range of voice signals is mainly 300–3,400 Hz with the propagation speed of $c = 3 \times 10^8$ m/s in free space. So, the wavelength of voice signals can be estimated according to eq. (1.1):

$$\lambda = c/f \tag{1.1}$$

The required antenna should be as large as several hundred kilometers, which is a huge demand for space and materials. If we assume signals are in much broader ranges, antennas, hence, are expected to be in all different sizes in order to work for them all and it is not clever to act this way. Modulation is the solution to this problem which shifts low-frequency spectrum up to the neighborhood of certain high-frequency carrier wave. As a result, antennas can be designed to be very compact in size and adaptive for broader low frequencies.

Modulation is the process of regulating one or more of the parameters of high-frequency carrier waves by modulation signals. The parameter could be amplitude, frequency or phase, which varies in accordance (usually linearly) with voltage of the modulation signals. Depending on the parameter being adjusted, modulation can be amplitude modulation (AM), frequency modulation (FM) or phase modulation. After modulation, carrier wave becomes a so-called modulated signal carrying the information of low FM signals.

Antenna size can be significantly reduced after employing modulation technique. If the antenna size is one-fourth of radiation signal wavelength, it will be more convenient to improve the radiation capability.

Civil AM radio frequency is 535–1,605 kHz (intermediate frequency, IF) with the corresponding wavelength of 187–560 m. The antenna size is about dozens of meters or hundreds of meters. Mobile phone uses 900 MHz band, and its antenna is only about 10 cm. The spectrum of the broadcast and voice signals is, therefore, shifted to required bands by modulation. Besides, modulation can also distribute information to different carrier waves, so that the receiver can choose to receive the specific information by its carrier wave frequency.

We take the AM broadcasting transmitting system as an example to illustrate the transmitting unit. There are three main parts: high-frequency part, low-frequency part and power. Its composition diagram and waveform at each stage are shown in Fig. 1.3.

High-frequency part consists of main oscillator, buffer stage, frequency multiplier, high-frequency amplifier and high-frequency power amplifier [1]. Each block in Fig. 1.3 is explained as follows.

(1) Main oscillator. Main oscillator provides a highly stable high-frequency oscillation signal f_{osc}, which is generally more than dozens of kHz.

(2) Buffer stage. Buffer stage is usually an amplifier circuit, which weakens the influence of the next stage back on main oscillator to ensure its frequency stability.

(3) Frequency multiplier. Frequency multiplier is composed of one or multiple stages of resonant amplifiers. It increases f_{osc} to the required carrier wave frequency f_c.

(4) High-frequency amplifier. High-frequency amplifier provides large enough carrier wave power.

(5) High-frequency power amplifier. The final power amplifier is high-frequency power amplifier, which is also used to provide sufficient transmitting power for signals.

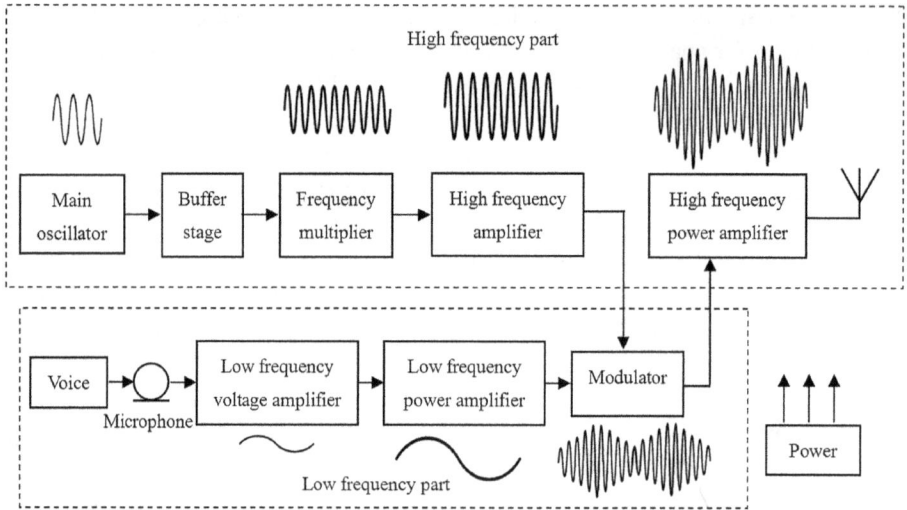

Fig. 1.3: Diagram of the AM transmitting system.

Low-frequency part of this transmitting system includes low-frequency voltage amplifier, low-frequency power amplifier and modulator.

(1) Low-frequency amplifier. Low-frequency amplifier is composed of multistage small-signal amplifiers, which is used to amplify electrical signals collected from the microphone.

(2) Modulator. Modulator is also called the final modulation low-frequency power amplifier, which is amplitude modulation amplifier in this diagram. Amplitude modulation amplifier combines carrier waves and the modulation signals into modulated signals.

1.2.2 Channel

EM waves radiated from antenna will then propagate through the channel. Different frequencies have different propagation modes. The common propagation modes are ground wave, sky wave and space wave [2].

1.2.2.1 Ground wave

Radio waves can travel along the curved surface of earth as shown in Fig. 1.4(a). The propagation of ground wave is relatively stable since the electrical property of earth rarely changes much in time. However, this transmission mode is limited by the skin effect. When there is an alternating current or alternating EM field in the conductor, the distribution of current is nonuniform. Most of the current will be concentrated on

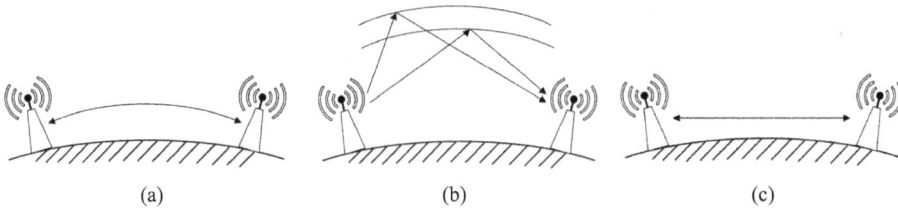

Fig. 1.4: Propagation modes of radio: (a) ground wave; (b) sky wave; and (c) space wave.

the "skin" of the conductor, the closer to the surface of the conductor, the greater the current density, the smaller the actual current inside the wire. As a result, the resistance of the conductor surface will increase, as well as the loss power. Earth is a good conductor, the higher frequencies, the greater of the skin effect and the more of the energy loss. Ground wave propagation is suitable for the medium wave and short-wave with the wavelength more than 200 m (frequency <1,500 kHz). The propagation distance of ground waves is not very far and typically in the range of several hundred kilometers.

1.2.2.2 Sky wave
From 50 km to several hundred kilometers above the ground, air molecules in the atmosphere ionize because of sunshine, producing sufficient positively charged atoms and free electrons. This atmosphere area is called ionosphere. Transmission of sky wave depends mainly on ionosphere reflection shown in Fig. 1.4(b). It is suitable for waves whose wavelength ranges from 10 to 200 m (frequency from 1,500 kHz to 30 MHz). Microwave with less than 10 m wavelength will pass through the ionosphere and escape; long wave with more than 3,000 m wavelength will be absorbed mostly by the ionosphere. As to the medium and shortwave signals, the shorter wavelength the stronger reflection we obtain. Therefore, sky wave mode is the most suitable for short-wave propagation and the signals can be reflected to thousands of kilometers away. However, the ionosphere is unstable and the sky wave will be absorbed and attenuated by ionosphere too. The greater ionization concentration, the greater power loss occurs. The fluctuation of EM waves caused by random variation of the ionosphere is called fading phenomenon. Ionization is high during the day and low at night. During night-time the absorption of intermediate waves reduces, which explains why we can receive more medium and intermediate waves broadcasting at night.

1.2.2.3 Space wave
Microwave bands refer to waves of less than 10 m wavelength (or higher than 30 MHz frequency). Microwave and ultra-shortwaves cannot be transmitted in forms of ground wave or sky wave due to the limitation of these two propagation modes. Instead they are more often transmitted in air along a straight path. This

mode is called space wave as shown in Fig. 1.4(c). Straight-line transmission encounters less atmosphere interference and less energy loss, so signals are strong and stable. TV and radar are both microwave signals. Due to the spherical earth surface, transmitting and receiving antennas need to be built very high if microwave travels without touching the ground. In order to increase propagation coverage as well as avoiding building high antennas, relays or base stations are needed. First, microwaves are launched from the origin and received by the relay station. After being amplified and sent to the next relay one after another, signals can travel a long distance to its destination like a relay race. For instance, when the synchronous communication satellite, 36,000 km high above the equator, is used to transmit microwaves, signals are able to cross the continent and ocean by using the satellite as the relay station.

1.2.3 Receiving equipment

Receiving equipment is mainly composed by receiving antenna, receiver and transducer. The block diagram is shown in Fig. 1.5.

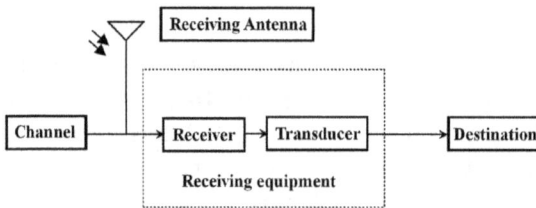

Fig. 1.5: Diagram of receiving equipment.

Receiving is the inverse process of transmitting. After long distance of transmission, EM waves have lost most of its power. Due to EM noises from various industrial equipments and interference of the atmosphere and the universe, receiving equipment takes in both useful signals and tremendous unwanted signals. So, receiving equipment is required to pick up faint useful signals out of the noisy EM background.

Receiving antenna picks up the EM wave in space and turns it into high-frequency electric oscillation.

Next, receiver converts the electric oscillation into electric signals, amplifies them selectively and obtains useful ones. This process is called demodulation. Demodulation is to recover modulation signals (low frequency) from the modulated signals, which is the inverse process of modulation.

There are numerous types of receivers such as superheterodyne, mirror frequency suppression, direct transformation, zero IF, digital IF and so on. The most popular radio receiver at present is the superheterodyne-type receiver. Superheterodyne receiver mixes local oscillator frequency with the input frequency reproducing modulation signal. The upmost characteristic of superheterodyne receiver is using a fixed-frequency IF amplifier to complete frequency selection and amplification. When the incoming frequency varies, local oscillator frequency also adjusts accordingly to maintain the IF unchanged. This approach is well liked for easy adjustment, good selectivity and high-frequency application. The disadvantages too are clear though. Superheterodyne receiver circuits are complex and allow combination frequency interference, subchannel interference, intermodulation interference and so on. So far, the superheterodyne receiver is still the mainstream technique in remote signal reception and has been applied to many measurement technologies.

Transducer restores electrical signal back to original information for the terminal user.

Let us take the radio broadcasting with superheterodyne receiver system as an example. A weak AM signal is received from antenna with carrier frequency of f_c. After amplification by high-frequency small-signal amplifiers, it is mixed with local oscillation signal with frequency f_L. The output is an AM signal with a different carrier frequency f_I but the same envelope. f_I is called the IF signal, which is the difference or sum of f_L and f_c. Being amplified by IF amplifier, signal enters the demodulation circuit. Low FM signal is extracted. Finally, modulation signal is amplified by low-frequency amplifiers and sent to speaker. The block diagrams and waveforms at each stage are displayed in Fig. 1.6.

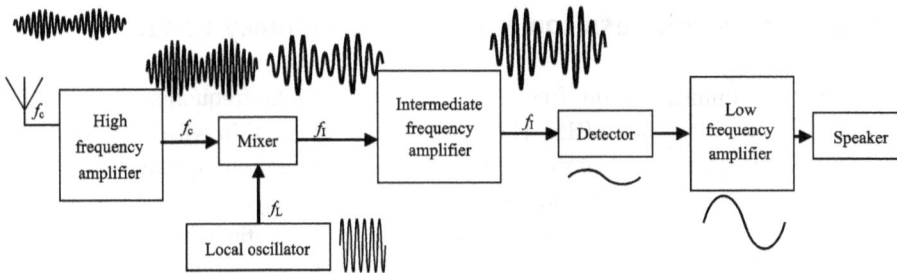

Fig. 1.6: Compositions diagram of superheterodyne receiving system.

Each block in Fig. 1.6 is explained as follows.
(1) High-frequency amplifier. High-frequency amplifier consists of one or several stages of small signal resonant amplifiers. It amplifies faint signals received from antenna with selected frequency range and filtering. The load is usually LC resonant circuits to emphasize frequency selection. The resonant frequency of high-frequency amplifier is identical to the carrier wave frequency f_c.
(2) Local oscillator. Local oscillator is used to generate local oscillation signal f_L. The frequency of f_L can be adjusted to track f_c.

(3) Mixer. Mixer is the core part of superheterodyne type receiver. There are two input signals of mixer, one being the high-frequency modulated signal with carrier frequency f_c and the other being the local oscillation signal with frequency f_L. The frequency of modulated signal will be converted to a fixed IF f_I through mixer, which is called the heterodyne effect. When the frequency of modulated signal changes, the local oscillator's frequency will change accordingly to keep the IF fixed. The expression of f_I is as follows:

$$f_I = |f_L - f_c| \qquad (1.2)$$

In China, the IF for AM radio is f_I = 465 kHz; for FM radio is f_I = 10.7 MHz; for TV is f_I = 38 MHz; for microwave receiver and satellite receiver are 70 and 140 MHz, respectively.

(4) IF amplifier. IF amplifier is also composed by small signal resonant amplifiers. Since the IF stays unchanged by adjusting f_L automatically with f_c, selectivity and gain of the IF amplifier are independent of the received carrier frequency.

(5) Detector. Detector realizes demodulation of signals, which extracts the low FM signal out of IF-modulated wave.

(6) Low-frequency amplifier. Low-frequency amplifier includes small-signal amplifiers and power amplifiers. It amplifies the useful modulation signals and provides enough power to drive the speaker.

1.3 Division of wireless communication frequency bands

In the field of communication, frequency bands refer to the frequency range of EM waves with the unit of hertz (Hz). The frequency bands used in wireless communication only occupy small portions of EM wave bands. Classified by wavelength, EM waves include ultra-long wave, long wave, medium wave, shortwave, ultra-shortwave (meter wave) and microwave (including decimeter wave, centimeter wave and millimeter wave). By frequency, there are very-low frequency, low frequency, IF, high frequency, very-high frequency, ultra-high frequency (UHF), super-high frequency and extremely high frequency.

All the radio signals are transmitted through air or free space. In order to share the spectrum reasonably and avoid frequency overlap of industrial, business or military sources, international telecommunications union-radio (ITU-R) Communications sector promulgates the international radio regulations which unifies the usage and range of radio frequency bands. The frequency divisions are slightly different in real practices from place to place, but all performances must be under regulation. In accordance with provisions of the international radio regulations, the existing radio communications are divided into more than 50 different types such as air communication, navigation communication, terrestrial communication, satellite communication, radio, television, radio navigation, positioning and telemetering, remote control, space exploration, and so on. Each application is stipulated within a certain frequency band. Some commonly used frequency bands in our country are as follows [3]:

Table 1.1: Common used frequency bands.

Application	Frequency bands
Medium wave AM broadcasting	535–1605 kHz
Shortwave AM broadcasting	230 MHz
FM broadcasting	88108 MHz
TV channels	50–100 MHz (1st–5th channels)
	170–220 MHz (6th–12th channels)
	470–870 MHz (beyond 13th channels)
Satellite direct TV (SDTV)	4–6 GHz; 12–14 GHz
Satellite direct broadcasting (SDB)	12–14 GHz
Global satellite positioning system (GPS)	L1 = 1,575.42 MHz; L2 = 1,227.60 MHz

Table 1.2 lists the divisions of radio frequency bands and the main applications. The frequency bands listed are not strictly divided and there is no absolute boundary between adjacent bands. "High frequency" is a relative concept. As long as the circuit size is much smaller than the operational wavelength, it can be considered as "high-frequency" range. The techniques of signal generation, amplification, transmission and receiving are different in different frequency bands. Usually signals with meter wave or longer wavelength are analyzed and generated by lumped parameters and the concept of "path." Signals with wavelength less than meter wave are analyzed and generated by distribution parameters and the concept of "field." In this book, the discussed frequency range covers from IF to UHF.

Currently, the commonly used wireless transmission frequencies are 2.4, 3.5, 5.8 GHz and so on. Frequencies of 2.4 and 5.8 GHz are open to public, and 3.5 GHz is the regional telecommunication band. These three frequencies have no diffraction ability but good reflection ability due to their short wavelengths.

Besides, 900, 700 and 400 MHz (UHF or VHF bands) have wavelength of about 1 m. Such signals are able to diffract, but the transmission bandwidth is narrow. Considering diffraction and transmission speed, vehicular mobile communication typically uses the bands below 800 MHz and above 450 MHz. A frequency of 350 MHz is the Public Security special band for police intercom. Paging bands are generally more than 130 and 150 MHz. A frequency of 230 MHz is mostly used for data transmission and 420–430 MHz is used for quasicluster band.

Table 1.2: Divisions of radio frequency (wave) bands.

Band	Abbreviation	Wavelength	Frequency range	Applications
Very long wave	VLF	100–10 km	3–30 kHz	Audio, telephone, data terminal, long-distance navigation, time scale, submarine navigation
Long wave	LF	10–1 km	30–300 kHz	Military underwater communications, power line communication, marine navigation, long-range navigation, time signals
Medium wave	MF	1000–100 m	300 kHz to 3 MHz	AM radio, amateur radio, mobile land communications, radio beacons
Shortwave	HF	100–10 m	3–30 MHz	Shortwave radio, amateur radio, mobile wireless telephone, fixed-point military communication
Very shortwave	VHF	10–1 m	30–300 MHz	FM radio, television, navigation, short-distance terrestrial, communication, mobile communications, air traffic control, vehicle, communications, wireless paging
Decimeter wave	UHF	1–0.1 m	300–3,000 MHz	TV, space telemetry, radar navigation, mobile communication, point-to-point communication
Centimeter wave	SHF	10–1 cm	3–30 GHz	Microwave communication, radar, wireless local area network, satellite links
Millimeter wave	EHF	10–1 mm	30–300 GHz	Communication, radar, microwave relay pass through the atmosphere, radio astronomy, research, military
Light wave		300–0.006 μm	1–50 THz	Optical fiber communication

Military radio frequency bands and devices are different from the civil uses. For example, military uses frequency hopping radios in which operating frequency changes from time to time. It is technically rare for general equipments to encounter or even capture these signals. Now, the bands of 1.4 and 1.2 GHz are increasingly preferred by military and other departments for mobile video transmission because of strong diffraction, clean band and good transmission effect. Table 1.3 is the radio frequency ranges named by the US military.

Table 1.3: Military radio frequency bands.

Band	Frequency range (GHz)	Wavelength (cm)
VHF	0.1–0.3	300–100
UHF	03–1.0	100–30
L	1.0–2.0	30–15
S	2.0–4.0	15–7.5
C	4.0–8.0	7.5–3.75
X	8.0–12	3.75–2.5
Ku	12–18	2.5–1.67
K	18–26	1.67–1.15
Ka	26–40	1.15–0.75
Millimeter wave	40–	0.75–

1.4 Frequency-selective circuits

In wireless communication, whether the system receives modulated signals from free space or realizes frequency conversion based on its nonlinear circuits, it requires the system to select useful signal out of complicated frequency components and reject useless interferences or noises. This is realized by frequency-selection circuits which are extremely important on signal quality and antiinterference ability. Frequency-selective circuits are the most basic network of high-frequency circuits, which are used widely in transmitting and receiving. Besides frequency selection, the circuits also apply for impedance-matching purpose in specific circumstances.

According to function, frequency-selective circuits can be divided into low-pass filter circuits, band-pass filter circuits, high-pass filter circuits and band-stop filter circuits. According to the operational principle, it can be divided into resonant circuits, crystal filters, ceramic filters, SAW (surface acoustic wave) filters and so on. Quartz crystal filters, ceramic filters and SAW filters are solid-state filters [4]. In the next part, we will introduce the principle and characteristics of various frequency-selective circuits.

1.5 *LC* resonant circuits

LC resonant circuits are composed of inductors (*L*) and capacitors (*C*). Based on the number of inductors and capacitors used, *LC* resonant circuits can be single-resonant circuits or dual-resonant circuits. In this chapter, single *LC* resonant circuits are the main topic, which are fundamental frequency-selective networks in nonlinear electronic circuits. There are so-called parallel *LC* resonant circuit and series *LC* resonant circuit depending on how *L* and *C* are arranged.

1.5.1 Equivalent conversion of series–parallel circuit

Equivalent conversion of series–parallel circuit is introduced first here. Series and parallel circuits both include resistance (R_S or R_P) and reactance (X_S or X_P) elements as shown in Fig. 1.7.

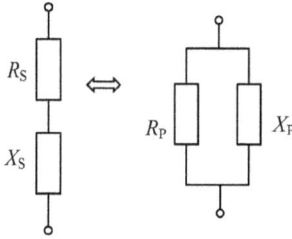

Fig. 1.7: Equivalent conversion of series–parallel circuit.

By assuming that these two circuits are equivalent, it requires identical impedance and frequency-selective property, which can be expressed by the following constraints:

$$\begin{cases} Z_P(j\omega) = Z_S(j\omega) \\ Q_P = Q_S \end{cases} \tag{1.3}$$

where Q is the quality factor reflecting frequency-selective performance, Z_P and Q_P represent impedance and quality factor of the parallel circuit, Z_S and Q_S represent impedance and quality factor of the series circuit, respectively. The expressions of impedance are therefore:

$$\begin{cases} Z_P(j\omega) = \frac{R_P \cdot jX_P}{R_P + jX_P} = \frac{X_P^2}{R_P^2 + X_P^2} R_P + j\frac{R_P^2}{R_P^2 + X_P^2} X_P \\ Z_S(j\omega) = R_S + jX_S \end{cases} \tag{1.4}$$

In order to satisfy the requirement of $Z_P(j\omega) = Z_S(j\omega)$, real parts and imaginary parts should match accordingly for both circuits:

$$\begin{cases} R_S = \frac{X_P^2}{R_P^2 + X_P^2} R_P \\ X_S = \frac{R_P^2}{R_P^2 + X_P^2} X_P \end{cases} \tag{1.5}$$

Or equivalently,

$$\begin{cases} R_P = \frac{R_S^2 + X_S^2}{R_S} \\ X_P = \frac{R_S^2 + X_S^2}{X_S} \end{cases} \tag{1.6}$$

The definitions about quality factor, Q, of series circuit and parallel resonant circuit are

$$Q_S = \frac{|X_S|}{R_S} = Q_P = \frac{R_P}{|X_P|} \tag{1.7}$$

According to eqs. (1.5)–(1.7), we obtain

$$\begin{cases} R_P = (1+Q^2)R_S \doteq Q^2 R_S \\ X_P = \left(1+\dfrac{1}{Q^2}\right)X_S \doteq X_S \end{cases} \tag{1.8}$$

Because $Q_P = Q_S$, we use Q to replace Q_P or Q_S in eq. (1.8). It is confirmed that the circuit properties during the series–parallel equivalent conversion are unchanged. Here reactance X_P and X_S can either be capacitance or inductance.

1.5.2 Basic characteristics of *LC* parallel resonant circuit

1.5.2.1 *LC* parallel resonant circuit without load
Basic structure of *LC* parallel resonant circuit without load is shown in Fig. 1.8, where r is the inherent loss resistance of inductor rather than load.

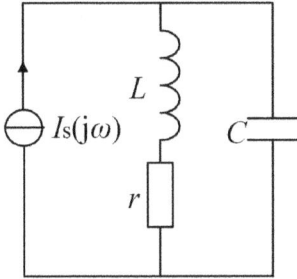

Fig. 1.8: Basic structure of *LC* parallel resonant circuit without load.

1.5.2.1.1 Resonance condition of the parallel resonant circuit
Admittance of the circuit in Fig. 1.8 is

$$Y(j\omega) = j\omega C + \frac{1}{j\omega L + r}$$

$$= \frac{r}{r^2 + (\omega L)^2} + j\left[\omega C - \frac{\omega L}{r^2 + (\omega L)^2}\right] \tag{1.9}$$

When $\omega L \gg r$, r becomes negligible and eq. (1.9) can be simplified as

$$Y(j\omega) = \frac{r}{(\omega L)^2} + j\left(\omega C - \frac{1}{\omega L}\right) \tag{1.10}$$

When $\omega C = \frac{1}{\omega L}$, susceptance (imaginary part of Y) is zero, which is the resonance condition. At this point, resonance angular frequency can be calculated $\omega = \omega_0 = \frac{1}{\sqrt{LC}}$ (or resonance frequency $f = f_0 = \frac{1}{2\pi\sqrt{LC}}$).

Also at resonance, $Y(j\omega) = \frac{r}{(\omega_0 L)^2} = G_0$ is a pure conductance, which is called the resonant conductance. Resonant resistance R_0 is

$$R_0 = \frac{1}{G_0} = \frac{(\omega_0 L)^2}{r} = Q_0 \omega_0 L = \frac{Q_0}{\omega_0 C} = Q_0{}^2 r = \frac{L}{Cr} \tag{1.11}$$

where $Q_0 = \frac{\omega_0 L}{r} = \frac{1}{\omega_0 Cr} = \frac{R_0}{\omega_0 L} = R_0 \omega_0 C$ is an important parameter known as the quality factor, subscript 0 in Q_0 means without load. Q factor is defined as the ratio between stored energy relative to energy loss of resonator. It reflects the frequency selectiveness and suppression ability of interference signal. Higher Q suggests better selectivity and lower energy loss. Q_0 is usually less than 200.

Equation (1.11) illustrates that when parallel resonant circuit is at resonance, the circuit is purely resistive. Another way of representing Fig. 1.8 is shown in Fig. 1.9 where internal property r is included in the total effective resonant resistance R_0. In the following paragraphs, we will use Fig. 1.9 to discuss LC parallel resonant circuit further. In fact, it is an equivalent series-to-parallel conversion from Figs. 1.8–1.50.

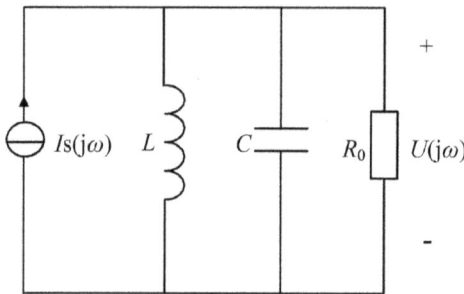

Fig. 1.9: *LC* parallel resonant circuit.

1.5.2.1.2 Resonant characteristics
Under resonant condition, there are several characteristics of LC parallel circuit that is worth noting.
(1) Susceptance is zero and admittance $|Y(j\omega)| = G_0$ reaches its minimum. The circuit is purely resistive.
(2) Voltage amplitude reaches maximum and $U(j\omega)$ is in phase with $I_S(j\omega)$.

$$U(j\omega) = U_0(j\omega) = I_S(j\omega) \cdot R_0 \tag{1.12}$$

(3) Branch current of capacitor is

$$I_{CO}(j\omega) = \frac{U_0}{\frac{1}{j\omega_0 C}} = j\omega_0 C U_0 = j\omega_0 C I_S(j\omega) R_0 = j\omega_0 C I_S(j\omega) \frac{Q_0}{\omega_0 C} = jQ_0 I_S(j\omega) \qquad (1.13)$$

Branch current of inductor is

$$I_{LO}(j\omega) = \frac{U_0}{j\omega_0 L} = \frac{I_S(j\omega) R_0}{j\omega_0 L} = \frac{I_S(j\omega) Q_0 \omega_0 L}{j\omega_0 L} = -jQ_0 I_S(j\omega) \qquad (1.14)$$

So,

$$|I_{CO}(j\omega)| = |I_{LO}(j\omega)| = Q_0 |I_S(j\omega)| \qquad (1.15)$$

It can be observed that I_{CO} and I_{LO} are equal in magnitude but opposite in phase. The current magnitude both equals the product of Q_0 and the current of the source.

(4) As to the off-resonance state, when the frequency of the signal source is higher than the resonant frequency f_0, the circuit is capacitive; when the frequency of the signal source is lower, the circuit is inductive.

1.5.2.1.3 Impedance characteristic curves

Impedance of *LC* parallel resonant circuit is

$$Z_P(j\omega) = \frac{1}{Y(j\omega)} = \frac{1}{G_0 + j(\omega C - \frac{1}{\omega L})}$$
$$= \frac{R_0}{1 + j\frac{(\omega C - \frac{1}{\omega L})}{G_0}} \qquad (1.16)$$

where

$$\frac{(\omega C - \frac{1}{\omega L})}{G_0} = \frac{\omega C \omega_0 L - \frac{\omega_0}{\omega}}{G_0 \omega_0 L} = Q_0 \left(\frac{\omega}{\omega_0} - \frac{\omega_0}{\omega} \right) \qquad (1.17)$$

When the detuning parameter $\Delta\omega = \omega - \omega_0$ is very small meaning ω is very close to ω_0, we can approximate the term as

$$\frac{\omega}{\omega_0} - \frac{\omega_0}{\omega} = \frac{(\omega + \omega_0)(\omega - \omega_0)}{\omega_0 \omega} \approx \frac{2\Delta\omega}{\omega_0} = \frac{2\Delta f}{f_0} \qquad (1.18)$$

Hence, eq. (1.16) can be approximated as

$$Z_P(j\omega) = \frac{R_0}{1 + jQ_0 \left(\frac{\omega}{\omega_0} - \frac{\omega_0}{\omega} \right)} = \frac{R_0}{1 + jQ_0 \left(\frac{f}{f_0} - \frac{f_0}{f} \right)} \approx \frac{R_0}{1 + jQ_0 \frac{2\Delta\omega}{\omega_0}} = \frac{R_0}{1 + jQ_0 \frac{2\Delta f}{f_0}} \qquad (1.19)$$

where $Q_0 \frac{2\Delta\omega}{\omega_0}$ is called the general detuning parameter. The norm of impedance is then

$$|Z_P| = \frac{R_0}{\sqrt{1+Q_0^2\left(\frac{\omega}{\omega_0}-\frac{\omega_0}{\omega}\right)^2}} = \frac{R_0}{\sqrt{1+Q_0^2\left(\frac{f}{f_0}-\frac{f_0}{f}\right)^2}} = \frac{R_0}{\sqrt{1+\left(Q_0\frac{2\Delta f}{f_0}\right)^2}} \tag{1.20}$$

The phase of impedance is

$$\varphi_Z = -\arctan\frac{\omega C - \frac{1}{\omega L}}{G_0} \tag{1.21}$$

From eq. (1.20) the amplitude–frequency and the phase–frequency characteristics of impedance of a parallel resonant circuit are shown in Fig. 1.10.

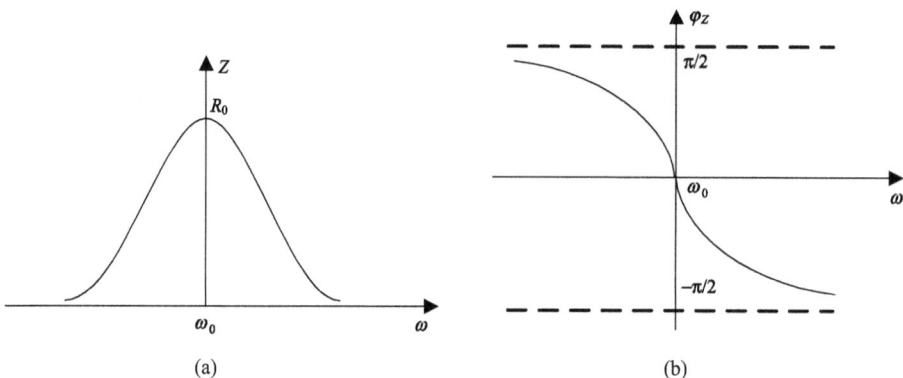

Fig. 1.10: Frequency characteristic of LC parallel resonant circuit, (a) impedance-frequency characteristic, (b) phase-frequency characteristic.

The plots suggest that when the operating frequency equals resonant frequency ω_0, $|Z|$ reaches its maximum value R_0 and the output voltage is the strongest at this moment. Once the frequency deviates from ω_0, $|Z|$ and the output voltage both reduces. Frequency selectiveness can be realized based on this property.

1.5.2.1.4 Frequency-selective characteristics of *LC* parallel resonant circuit
The voltage across *LC* parallel resonant circuit is

$$U(j\omega) = I_S(j\omega)Z_P(j\omega) \tag{1.22}$$

By substituting eq. (1.20), the amplitude–frequency characteristic expression is obtained

$$U = I_S \cdot |Z_P| = \frac{I_S \cdot R_0}{\sqrt{1 + Q_0^2 \left(\frac{f}{f_0} - \frac{f_0}{f}\right)^2}} = \frac{U_0}{\sqrt{1 + Q_0^2 \left(\frac{f}{f_0} - \frac{f_0}{f}\right)^2}} = \frac{U_0}{\sqrt{1 + \left(Q_0 \frac{2\Delta f}{f_0}\right)^2}} \tag{1.23}$$

The phase–frequency characteristic expression is

$$\varphi_U = -\arctan Q_0 \left(\frac{f}{f_0} - \frac{f_0}{f}\right) \tag{1.24}$$

From eq. (1.23) we can tell that voltage characteristic depends on the detuning parameter, the resonant frequency and also Q_0. A larger Q_0 makes both the amplitude–frequency and the phase–frequency characteristic curves look compressed horizontally with steeper slopes.

    The characteristic curves with different Q values ($Q_1 < Q_2$) are displayed in Fig. 1.11.

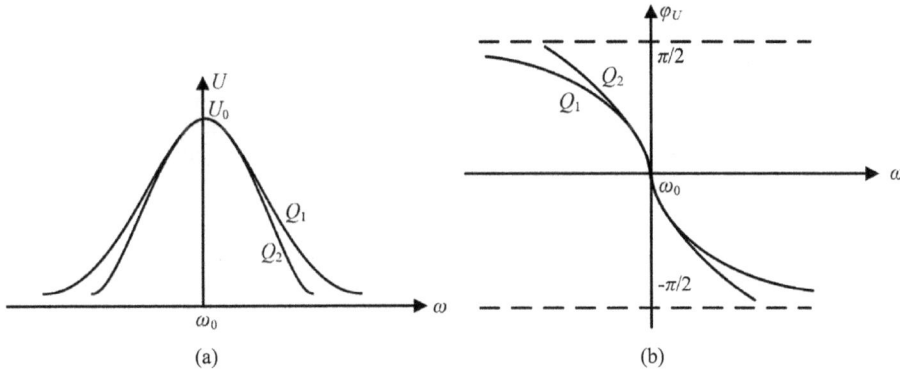

Fig. 1.11: (a) Amplitude–frequency, (b) phase–frequency characteristic curves with different Q values.

Normalized voltage is used to represent the amplitude–frequency characteristic expression:

$$|\alpha| = \frac{U}{U_0} = \frac{1}{\sqrt{1 + Q_0^2 \left(\frac{f}{f_0} - \frac{f_0}{f}\right)^2}} = \frac{1}{\sqrt{1 + \left(Q_0 \frac{2\Delta f}{f_0}\right)^2}} \tag{1.25}$$

It describes the selectivity of circuit. Smaller α represents better frequency selectivity. Normalized characteristic curve has the similar shape to Fig. 1.11 with different vertical readings (maximum value is 1).

1.5.2.1.5 Bandwidth

Bandwidth is also called passband. It usually refers to the frequency range within which signals are higher than $1/\sqrt{2}$ (or 3 dB drop) of the central maximum magnitude. As illustrated in Fig. 1.12 the bandwidth is noted in this graph as $B_{0.7}$:

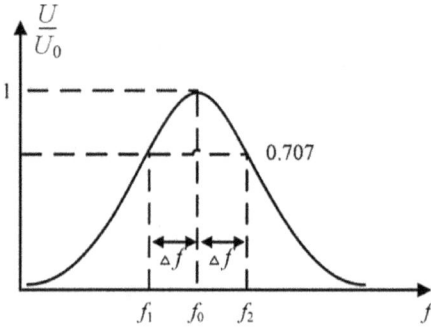

Fig. 1.12: Bandwidth of LC parallel resonant circuit.

$$B_{0.7} = f_2 - f_1 \tag{1.26}$$

We can find out the relationships between bandwidth $B_{0.7}$ and quality factor Q_0. According to the definition, frequencies at two ends of the bandwidth $(2\Delta f)$ satisfy

$$|\alpha| = \frac{U}{U_0} = \frac{1}{\sqrt{1 + \left(Q_0 \frac{2\Delta f}{f_0}\right)^2}} = \frac{1}{\sqrt{2}} \tag{1.27}$$

Solve eq. (1.27) for $2\Delta f$

$$Q_0 \frac{2\Delta f}{f_0} = \pm 1 \tag{1.28}$$

Assume that f_2 and f_1 are the upper- and lower-frequency ends, respectively, and rewrite eq. (1.28) as

$$\begin{cases} Q_0 \dfrac{2(f_2 - f_0)}{f_0} = 1 \\[2ex] Q_0 \dfrac{2(f_1 - f_0)}{f_0} = -1 \end{cases} \tag{1.29}$$

Substrate eq. (1.29)

$$Q_0 \frac{2(f_2 - f_1)}{f_0} = 2 \tag{1.30}$$

Therefore, the bandwidth is

$$B_{0.7} = f_2 - f_1 = \frac{f_0}{Q_0} \tag{1.31}$$

Equation (1.31) states that with a fixed f_0, large quality factor Q_0 results in small bandwidth $B_{0.7}$. Since quality factor and bandwidth values are related inversely, wide bandwidth and good selectivity can hardly be achieved both. Compromises usually need to be made when optimizing these two indices in designing an amplifier.

1.5.2.1.6 Rectangular coefficient

For an ideal resonant circuit, the amplitude–frequency curve is expected to be in a rectangular shape (Fig. 1.13) with the passband top being completely flat (signals within the bandwidth pass through with the same amplitude) and two tails being strict zero (signals outside the bandwidth are prohibited). The closeness to a rectangular shape of a curve can be defined to judge its selectivity, namely rectangular coefficient:

$$K_{0.1} = B_{0.1}/B_{0.7} \qquad (1.32)$$

where $B_{0.1}$ is the bandwidth allowing signals of more than one-tenth of the maximum voltage to pass through, parameter $K_{0.1}$ describes the difference between $B_{0.1}$ and $B_{0.7}$.

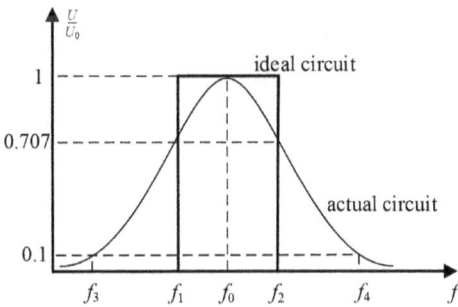

Fig. 1.13: Ideal circuit and actual circuit.

In the ideal scenario, $B_{0.1} = B_{0.7}$ and $K_{0.1} = 1$. However, for a curve with discussed amplitude–frequency property:

$$|\alpha| = \frac{U}{U_0} = \frac{1}{\sqrt{1 + \left(Q_0 \frac{2\Delta f}{f_0}\right)^2}} = \frac{1}{10} \qquad (1.33)$$

$B_{0.1}$ can be calculated as

$$B_{0.1} = \sqrt{10^2 - 1} \frac{f_0}{Q_0} \qquad (1.34)$$

$K_{0.1}$ is then

$$K_{0.1} = \sqrt{10^2 - 1} \doteq 9.95 \qquad (1.35)$$

$K_{0,1}$ is much larger than the ideal value of 1 and it is independent from quality factor or resonant frequency. This result demonstrates that the selectivity of single resonant LC circuit is relatively poor.

1.5.2.2 *LC* parallel resonant circuit with load
With load R_L and signal source resistor R_S connected, the LC parallel resonant circuit is plotted in Fig. 1.14.

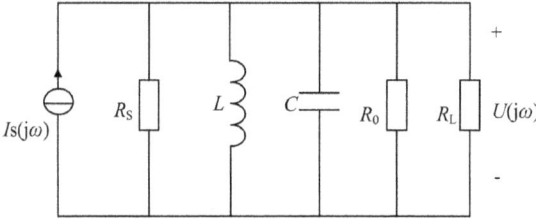

Fig. 1.14: *LC* parallel resonant circuit with signal source resistance and load.

Resonance condition remains $\omega_0 = \frac{1}{\sqrt{LC}}$. In order to distinguish unloaded and loaded status, Q_0 and Q_L are used to represent the quality factor with and without load.

When the circuit is unloaded,

$$Q_0 = \frac{R_0}{\omega_0 L} = R_0 \omega_0 C \tag{1.36}$$

When the circuit is loaded,

$$Q_L = \frac{R_\Sigma}{\omega_0 L} = R_\Sigma \omega_0 C = \frac{Q_0}{1 + \frac{R_0}{R_S} + \frac{R_0}{R_L}} < Q_0 \tag{1.37}$$

where $R_\Sigma = R_S // R_0 // R_L$.

Equation (1.37) shows that Q_L will increase as the source resistance and load resistance increases. Small R_S or R_L leads to diminished Q_L. It is obvious to conclude that high Q_L and good selectivity can only be maintained when R_S and R_L values are large enough. But no matter how we choose R_S and R_L, the quality factor is still more or less affected toward the lower end.

Besides, the signal source resistance does not necessarily equal that of the load, namely impedance mismatch. If the difference is significant, the load gains little power from the circuit. In actual cases, R_L and R_S are usually fixed and not adjustable. How to reduce the untoward effect of impedance mismatch? Impedance conversion will be a good solution here, where R_S and R_L are indirectly connected into LC resonant circuit with modified impedance. This method will be introduced in detail in the next section.

Other characteristics of loaded *LC* circuit are similar to unloaded case. All the equations mentioned in Section 1.5.2.1 are applicable in Section 1.5.2.2 too as long as Q_0 is replaced by Q_L. The following notations will not distinguish Q_0 and Q_L any longer. The audience needs to differentiate the actual meaning of Q depending on the context.

1.5.3 Basic characteristics of *LC* series resonant circuit

1.5.3.1 *LC* series resonant circuit without load

LC series resonant circuit without load is shown in Fig. 1.15, in which r is again the inherent loss resistance of inductor.

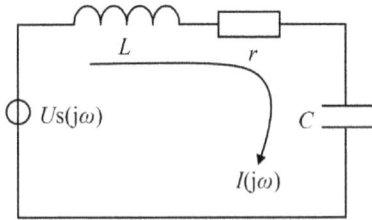

Fig. 1.15: *LC* series resonant circuit without load.

1.5.3.1.1 Resonance condition of series resonant circuit

In Fig. 1.15, impedance of series resonant circuit is

$$Z(j\omega) = r + j\omega L + \frac{1}{j\omega C} = r + j\left(\omega L - \frac{1}{\omega C}\right) \tag{1.38}$$

When $\omega L = \frac{1}{\omega C}$ and reactance of eq. (1.38) is zero, resonance condition is met. The resonant angular frequency is $\omega = \omega_0 = \frac{1}{\sqrt{LC}}$.

1.5.3.1.2 Resonant characteristics

When resonance occurs, there are several characteristics as follows:

(1) Reactance is zero. $|Z(j\omega_0)|$ is at its minimum value r and impedance is purely resistive.

(2) Loop current reaches the maximum,

$$I(j\omega) = I_0 = \frac{U_S(j\omega)}{r} \tag{1.39}$$

And $I(j\omega)$ is in phase with $U_S(j\omega)$.

(3) Voltage amplitude of inductor and capacitor is equal, which is Q_0 times of voltage source.

Quality factor is defined as

$$Q_0 = \frac{\omega_0 L}{r} = \frac{1}{r\omega_0 C} = \frac{1}{r}\sqrt{\frac{L}{C}} = \frac{\rho}{r} \tag{1.40}$$

where ρ is characteristic impedance with the value of

$$\rho = \omega_0 L = \frac{L}{\sqrt{LC}} = \sqrt{\frac{L}{C}} \tag{1.41}$$

Voltage on inductor is

$$U_{L0} = I_0 \cdot j\omega_0 L = \frac{U_S(j\omega_0)}{r} \cdot j\omega_0 L = j\frac{\omega_0 L}{r}U_S(j\omega_0) = jQ_0 U_S(j\omega_0) \tag{1.42}$$

Voltage on capacitor is

$$U_{C0} = \frac{I_0}{j\omega_0 C} = \frac{U_S(j\omega_0)}{r} \cdot \frac{1}{j\omega_0 C} = -j\frac{1}{\omega_0 Cr}U_S(j\omega_0) = -jQ_0 U_S(j\omega_0) \tag{1.43}$$

Therefore, we notice that

$$|U_{L0}| = |U_{C0}| = Q_0|U_S| \tag{1.44}$$

(4) In the off-resonance state, when operational frequency is higher than f_0, the circuit is inductive; when the operational frequency is lower than f_0, and the circuit is capacitive.

1.5.3.1.3 Energy relationships
Assuming the instantaneous loop current $i = I_m \sin \omega t$, the voltage on capacitor is calculated by the total charge accumulated over a certain time period t and its capacitance

$$u_C = \frac{1}{C}\int_0^t i \, dt = -\frac{I_{Cm}}{C\omega}\cos \omega t = -U_{Cm}\cos \omega t \tag{1.45}$$

Electrical energy of capacitor is

$$W_C = \frac{1}{2}Cu_C^2 = \frac{1}{2}CU_{Cm}^2\cos^2\omega t \tag{1.46}$$

Magnetic energy of inductor is

$$W_L = \frac{1}{2}Li^2 = \frac{1}{2}LI_m^2\sin^2\omega t \tag{1.47}$$

and

$$W_{\text{C max}} = \frac{1}{2}CU_{\text{Cm}}^2 = \frac{1}{2}CQ_0^2U_{\text{S}}^2 = \frac{1}{2}C\left(\frac{\rho}{r}\right)^2U_{\text{S}}^2 = \frac{1}{2}C\frac{1}{r^2}\frac{L}{C}U_{\text{S}}^2 = \frac{1}{2}LI_{\text{m}}^2 = W_{\text{Lmax}} \tag{1.48}$$

The total energy is

$$W = W_{\text{L}} + W_{\text{C}} = \frac{1}{2}LI_{\text{m}}^2 = \frac{1}{2}CU_{\text{Cm}}^2 = W_{\text{Lmax}} = W_{\text{Cmax}} \tag{1.49}$$

The total energy is a constant, which indicates that the energy flows between inductor and capacitor without additional loss.

1.5.3.1.4 Impedance characteristic curves

The impedance of *LC* series resonant circuit is

$$Z(j\omega) = r + j\left(\omega L - \frac{1}{\omega C}\right) \tag{1.50}$$

The amplitude–frequency characteristic equation and phase–frequency characteristic equation are, respectively,

$$|Z(j\omega)| = \sqrt{r^2 + \left(\omega L - \frac{1}{\omega C}\right)^2} \tag{1.51}$$

$$\varphi_Z(j\omega) = \arctan\frac{\omega L - \frac{1}{\omega C}}{r} \tag{1.52}$$

The curves are shown in Fig. 1.16. It can be concluded that the impedance magnitude is minimum at the resonant frequency ω_0. When the operational frequency is higher than ω_0, the circuit is inductive; when lower than ω_0, the circuit is capacitive.

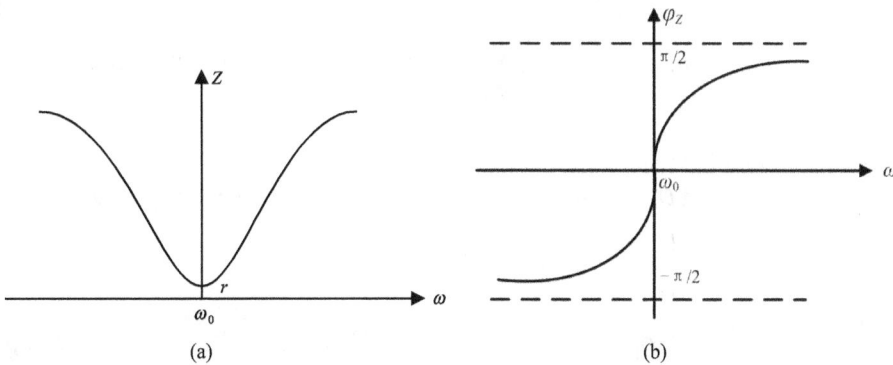

Fig. 1.16: Impedance characteristic curves of LC series resonant circuit: (a) amplitude–frequency and (b) phase–frequency.

1.5.3.1.5 Frequency-selective characteristics of *LC* series resonant circuit

The current of *LC* series resonant circuit is

$$I(j\omega) = \frac{U_S(j\omega)}{r + j\left(\omega L - \frac{1}{\omega C}\right)} \tag{1.53}$$

The normalized form is as follows:

$$\alpha = \frac{I(j\omega)}{I_0(j\omega)} = \frac{\dfrac{U_S(j\omega)}{r + j\left(\omega L - \frac{1}{\omega C}\right)}}{\dfrac{U_S(j\omega)}{r}} = \frac{1}{1 + j\left(\dfrac{\omega L - \frac{1}{\omega C}}{r}\right)} \tag{1.54}$$

$$= \frac{1}{1 + j\dfrac{\omega L}{r}\left(\dfrac{\omega}{\omega_0} - \dfrac{\omega_0}{\omega}\right)} = \frac{1}{1 + jQ_0\left(\dfrac{\omega}{\omega_0} - \dfrac{\omega_0}{\omega}\right)} = \frac{1}{1 + jQ_0\left(\dfrac{f}{f_0} - \dfrac{f_0}{f}\right)}$$

It is known from eq. (1.54) that the frequency characteristic of *LC* series resonant circuit has the same amplitude–frequency characteristic expression as with parallel circuit case shown in eq. (1.25). So, the characteristic parameters are the same for series and parallel circuits, such as the frequency-selective characteristics, band-width and rectangular coefficient.

1.5.3.2 *LC* series resonant circuit with load

A loaded *LC* series resonant circuit is shown in Fig. 1.17, with load R_L and signal source resistance R_S both considered.

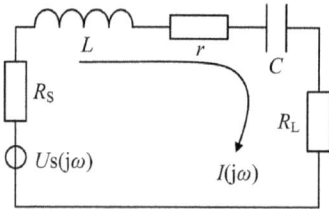

Fig. 1.17: *LC* series resonant circuit with the source resistance and load.

Resonance condition remains unchanged: $\omega_0 = \frac{1}{\sqrt{LC}}$.

Similarly, Q_0 and Q_L are used to distinguish unloaded and loaded quality factors.

$$Q_0 = \frac{\omega_0 L}{r} = \frac{1}{r\omega_0 C} \tag{1.55}$$

$$Q_L = \frac{\omega_0 L}{R_\Sigma} = \frac{1}{R_\Sigma \omega_0 C} = \frac{Q_0}{1 + \frac{R_S}{r} + \frac{R_L}{r}} < Q_0 \tag{1.56}$$

where $R_\Sigma = R_S + r + R_L$.

As eq. (1.56) uncovers, quality factor inevitably decreases once the signal source and load are connected to the LC resonant loop. In this way, the loaded circuit selectivity becomes poorer, but the bandwidth gets wider. As R_S and R_L increase, Q_L is further diminished.

Additionally, the actual signal source resistance and load are not purely resistive. When the frequency is very low, the reactance effect generally can be ignored. However, with the increase of operating frequency, the influence of reactance on resonant circuits should be taken into consideration.

Other characteristics are similar to our previous analysis of the unloaded circuit. All the mentioned equations in Section 1.5.3.1 (unloaded) are applicable for Section 1.5.3.2 (loaded) by only replacing Q_0 with Q_L.

1.6 Solid filters

Quartz crystal filters, ceramic filters and SAW filter are all solid filters. They do not need to be tuned. The operational principle and characteristics of solid filters will be discussed briefly in this section.

1.6.1 Quartz crystal filters

The chemical composition of quartz crystal is SiO_2, which is one of the most important electronic materials. Quartz crystal displays a special physical property, piezoelectric and inverse piezoelectric effect. When mechanical stresses are exerted on the specific quartz crystal surface, electric field will be induced which is proportional to the stress. This phenomenon is known as piezoelectric effect. On the other hand, when an electric field is applied on quartz crystal along a certain direction, mechanical vibrations are generated which are proportional to the field strength. It is inverse piezoelectric effect. Quartz crystal is able to convert mechanical vibrations into alternating electric field (voltage) and inversely. If the frequency of alternating voltage is equal to the natural frequency of quartz crystal, the crystal will resonate at its maximum mechanical vibration amplitude and the loop current also reaches the maximum. Besides piezoelectric effect, quartz crystal also has other desirable physical-stable and chemical-stable properties such as low temperature coefficient, hard texture, as well as its inactive, insoluble and nonflammable nature. Quartz crystal is an outstanding candidate material in applications of frequency selection and frequency stabilization.

The circuit symbol and equivalent electric model of quartz crystal are shown in Fig. 1.18.

The equivalent electric model is introduced to model the mechanical vibration performance into electrical properties. L_q is called the equivalent inductor or dynamic

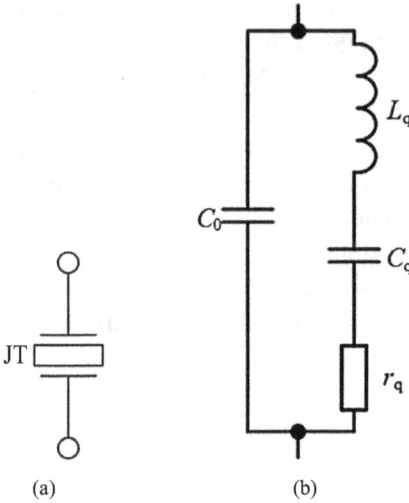

Fig. 1.18: (a) Circuit symbol, (b) equivalent electric model of quartz crystal.

(a) (b)

inductor and is related to the inertia of motion during vibrations. It is typically measured to be 10^{-3}–10^{-2} H. C_q is the equivalent capacitor or dynamic capacitor of 10^{-4}–10^{-1} pF. C_q is related to the elastic potential energy. r_q is the equivalent resistor or dynamic resistor roughly in the range of tens to hundreds of Ω. r_q stands for the mechanical loss during vibrations. C_0 is the static or distributive capacitor due to the coated electrodes, which is usually a few pFs.

As is noticed quartz crystal has a very small equivalent capacitance and resistance but a large equivalent inductance. Its quality factor $Q_q = \frac{1}{r_q}\sqrt{\frac{L_q}{C_q}} - 1$ is very large and can reach tens of thousands or even several millions, which is very much higher than an LC circuit built by common electrical components.

1.6.2 Ceramic filters

Some materials such as lead zirconate titanate $Pb(ZrTi)O_3$ can be made into ceramic filters. After direct current (DC) high-voltage polarization, ceramic also has piezo-electric effect just like quartz crystal.

A single ceramic is a 2-terminal device, whose symbol and equivalent circuit are the same with quartz crystal. The operational frequency range is from hundreds of kHz to several hundred of MHz. The bandwidth of ceramic filters is very narrow. The quality factor is about a few hundred, which is higher than LC filters and lower than quartz crystal filters. The frequency selectivity of 2-terminal ceramic filters is usually poor, so engineers use several ceramic filters together to form 4-terminal filters that perform better. The circuit and symbol of 4-terminal ceramic filters are as shown in Fig. 1.19.

Ceramic resonator

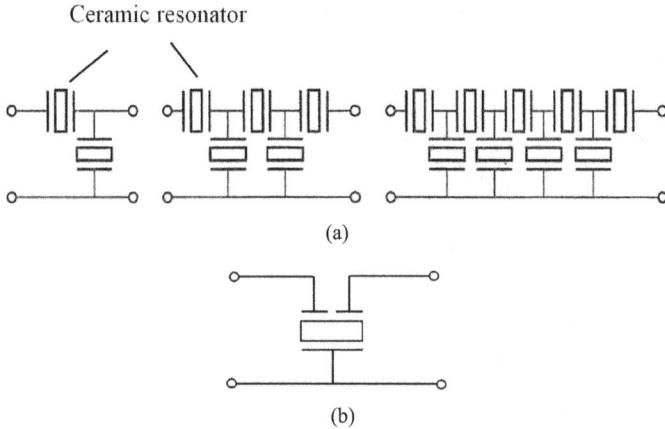

(a)

(b)

Fig. 1.19: (a) Four-terminal ceramic filters and (b) symbol.

Ceramic material is easy to be manufactured into arbitrary shapes and convenient in miniaturized production. It has excellent performance as well as low cost. Therefore, ceramic filters are widely used, such as the IF amplifiers in radio and TV receivers.

The ceramic filter IF amplifier circuit is shown in Fig. 1.20. Ceramic filter LB is parallel to emitter resistor R_e as a bypass capacitor. The resonant frequency of ceramic filter is equal to the IF. The more frequency shifts from the IF of the input, the larger impedance of the ceramic filter we get as well as the lower gain.

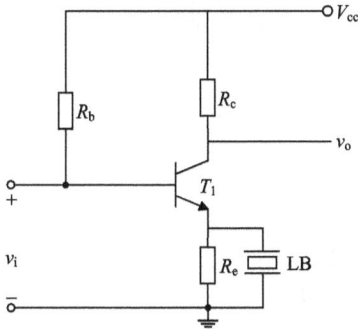

Fig. 1.20: Ceramic filter circuit.

1.6.3 Surface acoustic wave filters

SAW filters are made of materials with piezoelectric effect. There are interdigitated electrodes at both input and output terminals, which are called input interdigital transducer and output interdigital transducer. The SAW structure is shown in Fig. 1.21.

Interdigitated electrodes

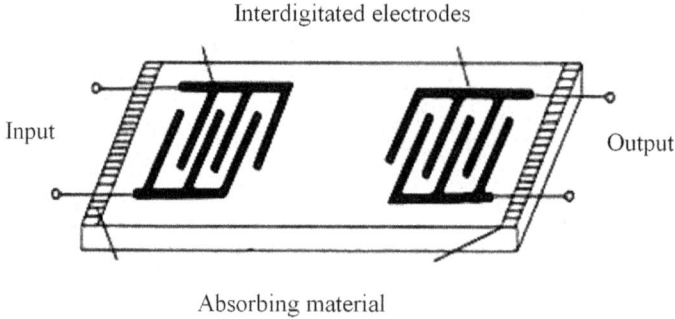

Absorbing material

Fig. 1.21: Schematic diagram of the SAW filter.

When an alternating current (AC) signal is applied to the transmitting transducer, the surface of piezoelectric substrate produces periodic vibrations. Parts of the acoustic wave signals are absorbed by the absorbing material and the rest is transmitted to the output.

When the frequency of input signal matches the natural frequency of SAW, the circuit has the highest transmission efficiency and lowest power loss. At the receiving terminal, the output interdigital transducer converts the acoustic wave into electrical signals.

The parameters such as center frequency, amplitude–frequency characteristic and phase–frequency characteristic of SAW filter is determined by the geometry of interdigital electrodes. The equivalent circuit and symbol are shown in Fig. 1.22. The input and output impedance is capacitive, mainly caused by the static capacitance of interdigital transducers. In real applications, an inductor and a resistor are often added parallel to the input and output to realize impedance matching.

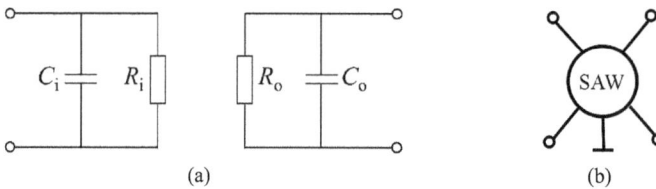

(a)   (b)

Fig. 1.22: (a) SAW filter equivalent circuit, (b) symbol.

SAW filters have advantages such as high operational frequency, broad bandwidth, good frequency selectivity and miniaturized sizes. The frequency bandwidth ratio $B_{0.7}/f_0$ is up to 50%. SAW filter shaves a nearly linear phase–frequency characteristic in the passband with a linear phase shift of $\pm 1.5°$, as shown in Fig. 1.23.

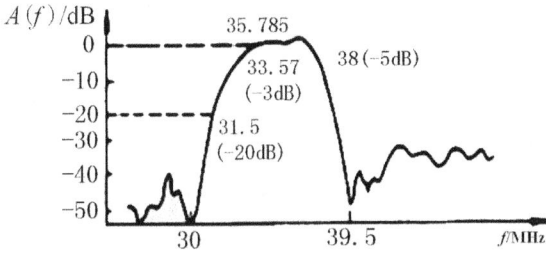

Fig. 1.23: SAW filter amplitude–frequency characteristics.

The manufacturing process of SAW filters is very simple and cost effective. Frequency consistency is nearly perfect. It is widely used in a variety of electronic equipments. In recent years, SAW filters have been integrated on-chip with the weight of only 0.2 g. In addition, thanks to the emerged new materials and the latest processing technology, the upper frequency of SAW filters is increased to 2.5–3 GHz. SAW filters, therefore, have wider applications in the field of Electromagnetic Interference (EMI) area.

1.7 Impedance conversion networks

In the communication electronic circuits, impedance conversion is often needed to match the signal source or load with the circuits. Impedance conversion refers to the process that the original impedance is modified to be the desired matching impedance. The maximum transmission power can be obtained when the impedance match is achieved. The transmitting efficiency and sensitivity of the receiver can also be improved.

Based on previous discussions, the resistance of signal source or load will degrade the quality factor Q and, hence, lower the selectivity of circuits. In order to reduce the influence from signal source and load on Q, impedance conversion is a popular solution for improving the performance of the whole circuits, especially the selectivity. In the following contents, some commonly used impedance conversion networks are introduced and analyzed.

1.7.1 Autotransformer circuit

Autotransformer impedance conversion network is shown in Fig. 1.24(a). The inductor coil is N_1 turns in total and the load R_L is tap-connected with N_2 turns. Figure 1.24(b) is the equivalent circuit of Fig. 1.24(a). R'_L is the equivalent resistance of R_L converted from the secondary circuit to the primary circuit while maintaining the same circuit performance.

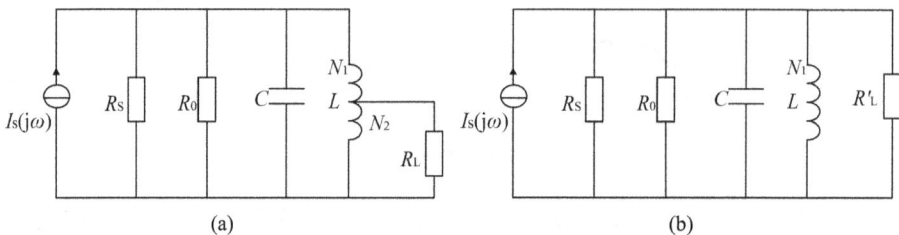

Fig. 1.24: Autotransformer impedance conversion network: (a) autotransformer circuit and (b) equivalent circuit of autotransformer.

Assuming the transformer is ideal, no magnetic leakage exists between the primary and the secondary coils. Power should be conserved before and after impedance conversion, $P_{primary} = P_{secondary}$. This can be expressed as

$$\frac{U^2_{primary}}{R'_L} = \frac{U^2_{secondary}}{R_L} \tag{1.57}$$

Voltage across the primary and the secondary coils depends on the turns of coils

$$\frac{U^2_{primary}}{U^2_{secondary}} = \left(\frac{N_1}{N_2}\right)^2 \tag{1.58}$$

Here, we need to define the access factor n, which describes the fraction of partially connected load voltage versus the total voltage. It reflects the effect of load on the main circuit. In the above autotransformer circuit, n can be calculated as

$$n = \frac{N_2}{N_1} \tag{1.59}$$

Equation (1.57) is rewritten as

$$R'_L = \frac{1}{n^2} R_L \tag{1.60}$$

By definition n is less than (partially connected) or equal to 1 (fully connected). After conversion, the equivalent R'_L is larger than R_L since R_L is tap-connected. Small access factor n associates with large equivalent load resistance. When a large load is connected there is little influence exerted on the main circuit. We can speculate this from the quality factor Q_L of LC parallel resonant circuit shown as below:

$$Q_L = \frac{R_\Sigma}{\omega_0 L} = \frac{R_0 // R_S // R'_L}{\omega_0 L} \tag{1.61}$$

R_L being partially connected to LC resonant circuit ($R'_L > R_L$) allows a greater quality factor Q_L compared to fully connected ($R'_L = R_L$).

Sometimes the load is not purely resistive and the reactance component needs to be taken into consideration (Fig. 1.25). The load is modeled as a resistor and a capacitor in parallel.

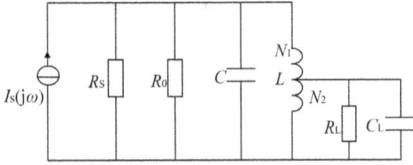

Fig. 1.25: Conversion network with capacitive load.

Here, the conversion relationship of R_L is the same as in eq. (1.60). As for capacitance C_L, similar procedures are performed

$$\frac{U^2_{\text{primary}}}{\frac{1}{\omega C'_L}} = \frac{U^2_{\text{secondary}}}{\frac{1}{\omega C_L}} \tag{1.62}$$

where C'_L is the equivalent capacitance after conversion

$$C'_L = \left(\frac{N_2}{N_1}\right)^2 C_L = n^2 C_L \tag{1.63}$$

After conversion, capacitance becomes smaller and the effect of load reactance on the circuit is reduced compared to the full connection case. So, the resonant frequency of circuit is less influenced by applying partial connection.

1.7.2 Mutual transformer circuit

Mutual transformer impedance conversion circuit is shown in Fig. 1.26 with primary coil of N_1 turns and the secondary coil of N_2 turns. Load R_L is connected to the secondary coil, Fig. 1.26(b) is the equivalent circuit after conversion.

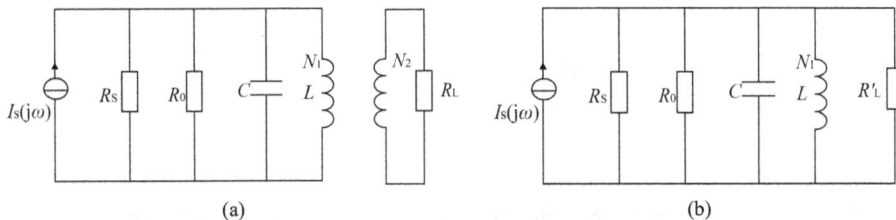

(a)

(b)

Fig. 1.26: Mutual transformer impedance conversion network: (a) transformer circuit and (b) equivalent circuit of a transformer.

Assuming the transformer is ideal and using the same analysis process of autotrans-former, we get $R'_L = \frac{1}{n^2}R_L$. If the load contains reactance component, eq. (1.63) is suitable in mutual transformer case too.

However, in practical cases the mutual transformer is not ideal. The transformer can be approximated as ideal as long as the coupling coefficient between the primary and secondary coils is close to 1.

1.7.3 Capacitor divider circuit

Capacitor divider impedance conversion circuit is shown in Fig. 1.27.

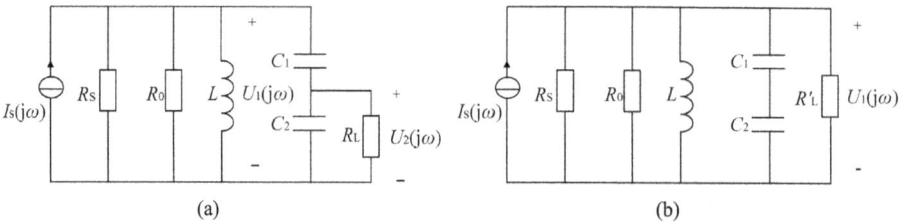

(a)   (b)

Fig. 1.27: Capacitor divider impedance conversion network: (a) capacitor divider circuit and (b) equivalent circuit of a capacitor divider.

Resonant capacitor is divided as C_1 and C_2. Load R_L is connected across C_2. The volt-age across the LC resonant circuit is $U_1(j\omega)$ and the acquired output voltage is $U_2(j\omega)$.

Assume the capacitor is ideal with no power loss. Similarly, the equivalence of conversion requires the same power consumption on the load before and after

$$\frac{U_1^2}{R'_L} = \frac{U_2^2}{R_L} \tag{1.64}$$

When $R_L \gg \frac{1}{\omega C_2}$ (the existence of R_L does not change the circuit impedance much), voltage U_1 and U_2 still satisfy

$$\frac{U_2}{U_1} = \frac{\frac{1}{\omega C_2}}{\frac{1}{\omega C_2} + \frac{1}{\omega C_1}} = \frac{C_1}{C_1 + C_2} \tag{1.65}$$

The access factor $n = \frac{C_1}{C_1 + C_2}$ and $R'_L = \frac{1}{n^2}R_L$.

If the load is capacitive, the previous conclusion $C'_L = n^2 C_L$ holds true here too.

1.7.4 Partial access of the signal source

In order to reduce the influence of signal source resistance on the circuit, partial access methodology can also be applied to the signal source. The procedure and analysis are almost identical given the symmetry of load and source on the main circuit.

Through the above deductions it can be found that quality factor Q_L is improved, resonant frequency is less affected and the circuit is more stable by impedance conversion. Moreover, by adjusting the value of access factor n, impedance matching between source and load can be achieved. However, all discussed partial access methods have nothing to do with the operational frequency. If impedance matching among the source, the circuit and load at a certain frequency is required, *LC* frequency matching networks should be considered which are covered in the next section.

1.8 *LC* frequency matching networks

A matching network is either input matching network or output matching network. Input matching network equalizes the output impedance of signal source with the input impedance of the amplifier circuit, so that the amplifier acquires maximum excitation power. Output matching networks match the load resistance with the output impedance of the amplifier, which ensures the output power is maximized. In addition, these matching networks can work as frequency selectors and filters.

Various types of frequency matching networks can be derived based on the technique of series–parallel circuit conversion which has been introduced in Section 1.5.1. The main structures are known as the L-type, T-type and π-type networks.

1.8.1 L-type matching network

L-type is the simplest frequency-selective matching network. In this section, we will explain L-type circuit in detail about the principle of series–parallel equivalent conversion.

L-type network contains two reactance components X_1 and X_2 with opposite properties. The commonly used circuits are shown in Fig. 1.28.

In both settings, R_2 is the actual load resistor and R_1 is the equivalent overall resistor tested from the left port at a certain frequency.

In Fig. 1.28(a), we assume X_1 is capacitor and X_2 is inductor (Fig. 1.29). X_1 and X_2 constitute the L-type matching network with the ability to increase the load.

X_2 and R_2 are connected in series. From the series–parallel conversion law, Fig. 1.29 is equivalent to the parallel structure shown in Fig. 1.30.

When in resonance, the circuit is purely resistive and the reactance part cancels out: $j(X_1 + X_P) = 0$ and $R_1 = R_P$.

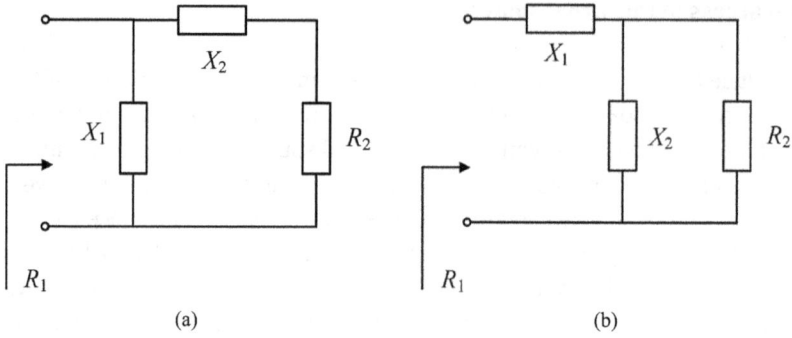

(a)   (b)

Fig. 1.28: L-type matching networks, (a) R1>R2, (b) R1<R2.

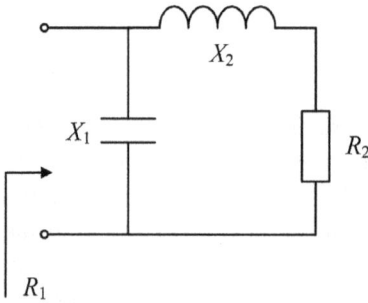

Fig. 1.29: L-type matching network to increase the load.

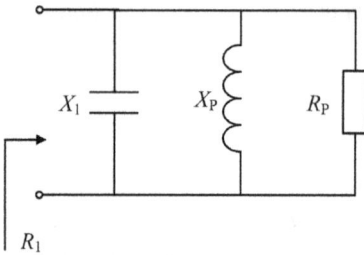

Fig. 1.30: Equivalent conversion of the L-type matching network for increasing load.

According to eq. (1.8)

$$R_1 = R_P = (1 + Q^2)R_2 > R_2 \qquad (1.66)$$

So, the equivalent load R_1 is larger than the actual load R_2. The L-type matching network increases the load.

From eq. (1.66) Q is calculated as

$$Q = \sqrt{\frac{R_1}{R_2} - 1} \qquad (1.67)$$

Impedance X_2 can be determined

$$|X_2| = QR_2 = \sqrt{R_2(R_1 - R_2)} \tag{1.68}$$

From eqs. (1.7) and (1.8) as well as the resonance condition X_1 can also be determined

$$|X_P| = |X_1| = \frac{R_P}{Q} = \frac{R_1}{Q} = R_1\sqrt{\frac{R_2}{R_1 - R_2}} \tag{1.69}$$

The anticipated capacitance and inductance are obtained from eqs. (1.68) and (1.69), respectively,

$$\begin{cases} L = \dfrac{|X_2|}{\omega} = \dfrac{|X_2|}{2\pi f} \\ C = \dfrac{1}{\omega|X_1|} = \dfrac{1}{2\pi f|X_1|} \end{cases} \tag{1.70}$$

All parameters above are functions of frequency. It suggests that different matching values are expected at different frequency points.

The other L-type circuit shown in Fig. 1.28(b) can be analyzed in the same way. Figure 1.31 illustrates the conversion process. Here, we assume X_1 is an inductor and X_2 is a capacitor. This L-type matching network is capable to decrease the load.

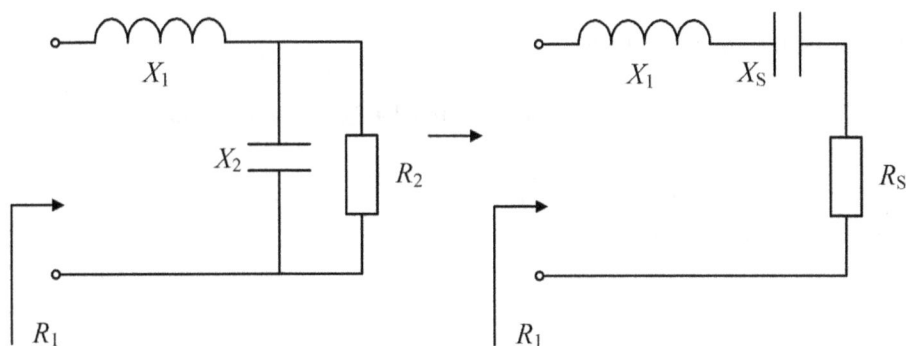

Fig. 1.31: L-type matching network for decreasing load.

Under resonance condition $j(X_1 + X_S) = 0$ and $R_1 = R_S$. From eq. (1.8), $R_1 = R_S = \frac{R_2}{1+Q^2} < R_2$ can be found. The equivalent load R_1 is smaller than the actual load R_2. The load is decreased.

Again, Q can be calculated as

$$Q = \sqrt{\frac{R_2}{R_1} - 1} \tag{1.71}$$

According to eqs. (1.7) and (1.8), both impedances X_1 and X_2 are determined:

$$|X_2| = \frac{R_2}{Q} = R_2\sqrt{\frac{R_1}{R_2 - R_1}} \tag{1.72}$$

$$|X_1| = |X_S| = QR_S = QR_1 = \sqrt{R_1(R_2 - R_1)} \tag{1.73}$$

Capacitance and inductance are accessible from eqs. (1.72) and (1.73)

$$\begin{cases} L = \dfrac{|X_1|}{\omega} = \dfrac{|X_1|}{2\pi f} \\ C = \dfrac{1}{\omega|X_2|} = \dfrac{1}{2\pi f|X_2|} \end{cases} \tag{1.74}$$

If the load is not a merely resistor R_2 but also includes reactance effect, the reactance can be merged as part of the L-type network in calculation. When the calculation is done, subtract the load reactance from the L-type network parameters to get the actual parameters.

From the earlier discussions, either of L-type matching network's parameters are closely related to quality factor Q. Q cannot be arbitrarily decided. Consequently, L-type network may not meet the requirements of filtering, which can be solved by T-type or π-type networks later.

1.8.2 T-type matching network

Three reactance components X_P, X_{S1} and X_{S2} build a T-type network shown in Fig. 1.32.

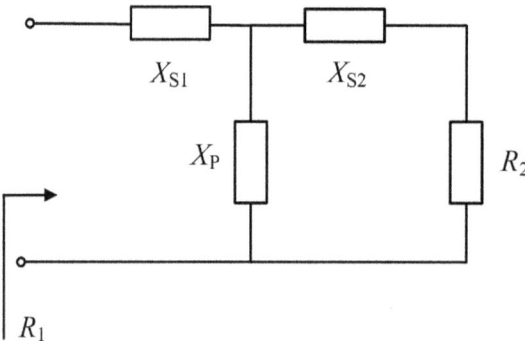

Fig. 1.32: T-type matching network.

T-type network can be seen as two L-type networks by splitting X_P into X_{P1} and X_{P2} (Fig. 1.33) with $X_P = X_{P1} // X_{P2}$. R' is the equivalent resistance of R_2 when the L-shaped X_{P2} and X_{S2} are added. R_1 is the equivalent resistance of R' when X_{P1} and X_{S1} are

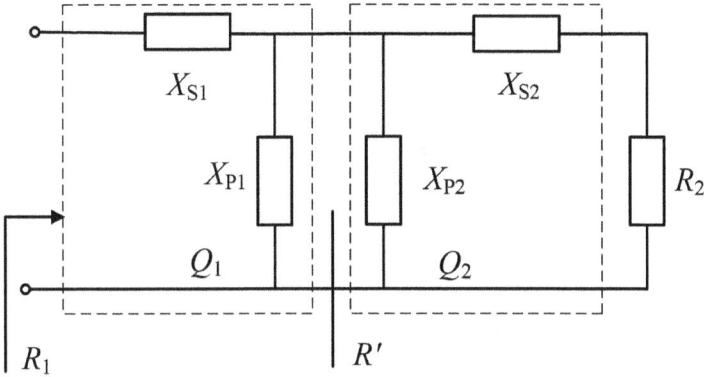

Fig. 1.33: Equivalent conversion of T-type matching network.

added. By thinking this way, we consider a T-type conversion is essentially a two-step L-type conversion.

Borrowing the conclusions of L-type network, X_{P2} and X_{S2} form an L-type matching network with

$$Q_2 = \sqrt{\frac{R'}{R_2} - 1} \qquad (1.75)$$

X_{S1} and X_{P1} form another L-type matching network with

$$Q_1 = \sqrt{\frac{R'}{R_1} - 1} \qquad (1.76)$$

The values of Q_1 and Q_2 can be determined based on the requirements of matching network as long as total Q satisfies $Q = Q_1 + Q_2$. Bandwidth of the network can be estimated by the value of Q. Reversely, the larger one of Q_1 and Q_2 can be determined based on bandwidth requirement. The determination of parameters is discussed in two separate scenarios.

(1) When $R_1 > R_2$ and $Q_2 > Q_1$, we can start the calculation from R_2. Q_2 must satisfy

$$Q_2 > \sqrt{\frac{R_1}{R_2} - 1} \qquad (1.77)$$

Then

$$
\begin{cases}
R' = (1 + Q_2^2)R_2 \\
X_{S2} = Q_2 R_2 \\
X_{P2} = \dfrac{R'}{Q_2} \\
X_{S1} = Q_1 R_1 \\
X_{P1} = \dfrac{R'}{Q_1}
\end{cases}
\tag{1.78}
$$

(2) When $R_2 > R_1$, $Q_1 > Q_2$, we start from R_1. Q_1 must satisfy

$$
Q_1 > \sqrt{\dfrac{R_2}{R_1} - 1}
\tag{1.79}
$$

Then

$$
\begin{cases}
R' = (1 + Q_1^2)R_1 \\
X_{S1} = Q_1 R_1 \\
X_{P1} = \dfrac{R'}{Q_1} \\
X_{S2} = Q_2 R_2 \\
X_{P2} = \dfrac{R'}{Q_2}
\end{cases}
\tag{1.80}
$$

1.8.3 π-Type matching network

The structure of a π-type matching network is shown in Fig. 1.34.

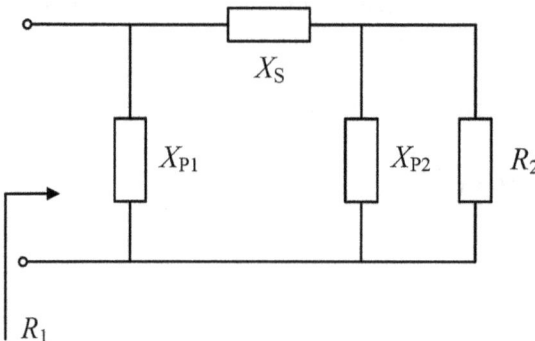

Fig. 1.34: π-Type matching network.

Similarly, a π-type matching network can also be decomposed into two L-type networks as shown in Fig. 1.35. X_S can be seen as the combination of X_{S1} and X_{S2} with $X_S = X_{S1} + X_{S2}$.

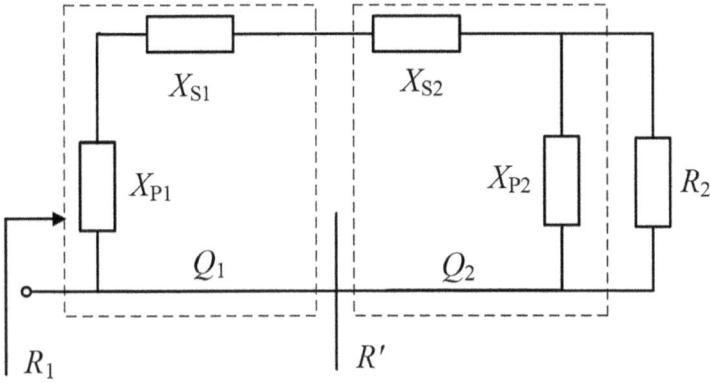

Fig. 1.35: Equivalent conversion of π-type matching network.

In the L-type network composed by X_{P2} and X_{S2},

$$Q_2 = \sqrt{\frac{R_2}{R'} - 1} \tag{1.81}$$

In the L-type network composed by X_{S1} and X_{P1},

$$Q_1 = \sqrt{\frac{R_1}{R'} - 1} \tag{1.82}$$

It is not difficult to see that $R' < R_1$ and $R' < R_2$ must be met to keep Q values rational. The values of Q_1 and Q_2 can be determined according to the performance of matching network. Parameters of the π-type network are also obtained in two scenarios.

(1) When $R_1 > R_2$ and $Q_1 > Q_2$, we can start the calculation from R_1. Select Q_1 which satisfies

$$Q_1 > \sqrt{\frac{R_1}{R_2} - 1} \tag{1.83}$$

Then

$$\begin{cases} R' = \dfrac{R_1}{(1 + Q_1^2)} \\ X_{S1} = Q_1 R' \\ X_{P1} = \dfrac{R_1}{Q_1} \\ X_{S2} = Q_2 R' \\ X_{P2} = \dfrac{R_2}{Q_2} \end{cases} \tag{1.84}$$

(2) When $R_2 > R_1$ and $Q_2 > Q_1$, we can start the calculation from R_2. Q_2 should satisfy

$$Q_2 > \sqrt{\frac{R_2}{R_1} - 1} \qquad (1.85)$$

Then

$$\begin{cases} R' = \dfrac{R_2}{(1 + Q_2^2)} \\ X_{S1} = Q_1 R' \\ X_{P1} = \dfrac{R_1}{Q_1} \\ X_{S2} = Q_2 R' \\ X_{P2} = \dfrac{R_2}{Q_2} \end{cases} \qquad (1.86)$$

1.9 Basic characteristics of nonlinear devices

1.9.1 Introduction

The main features of electronic circuits discussed in this course are high frequency and nonlinearity. The circuits are nonlinear as long as they contain one device with nonlinearity. In the communication electronic circuits, numerous devices are non-linear or let us say operating at nonlinear regions, such as diodes and transistors. Though the same devices are employed in analog circuits (low frequency) to build linear circuits, they manifest obvious nonlinearity under certain conditions. Instead of judging a device as strict linear or nonlinear, it is wiser to decide its property depending on the specific condition.

Nonlinear devices can be divided as nonlinear resistive device (Fig. 1.36(a)), non-linear capacitive device (Fig. 1.36(b)) and nonlinear inductive device (Fig. 1.36(c)).

Nonlinear resistors are with nonlinear current–voltage curves. Nonlinear capaci-tors are nonlinear in charge–voltage relationship. Nonlinear inductors refer to nonlin-ear magnetic flux and current relationship.

When analyzing the response of nonlinear devices, methods based on linear su-perposition law are no longer applicable. Nonlinear differential or integral equa-tions will be involved instead. The characteristic equations of nonlinear devices are different with different control variables. In real practices, reasonable approxima-tions can be made for simplification. Graphic method and analytical method are two commonly encountered approaches.

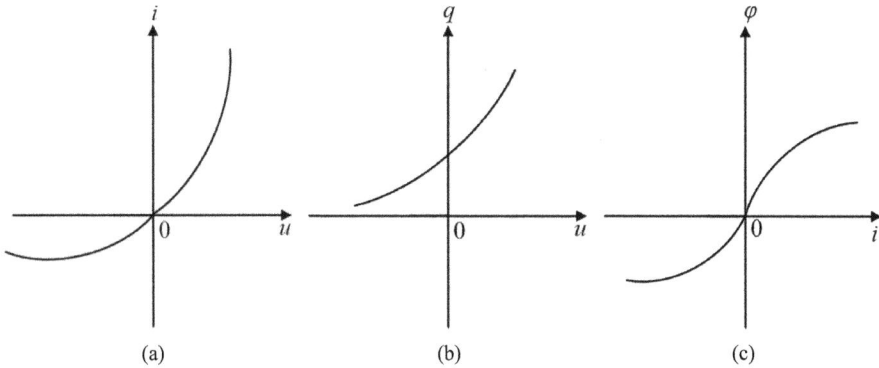

Fig. 1.36: Characteristics of Nonlinear device, (a) nonlinear resistive device, (b) nonlinear capacitive device, (c) nonlinear inductive device.

By graphic method, we obtain the output current and voltage waveforms by drawing. Analytical method makes use of mathematical expressions of nonlinear devices and solves for the solutions. These two methods will be introduced briefly.

1.9.2 Characteristics of nonlinear devices

Let us take the example of a nonlinear resistor device to illustrate the graphic method. Figures 1.37–1.39 show the i–u characteristic curves of this device, from which we can study its resistance or conductance property.

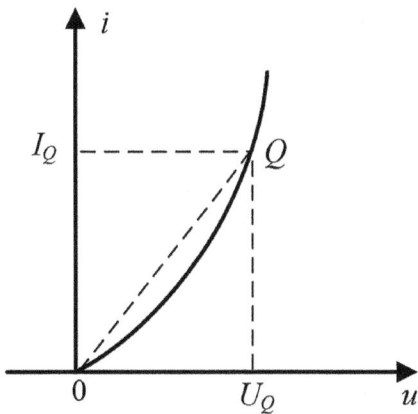

Fig. 1.37: Diagram of DC conductance.

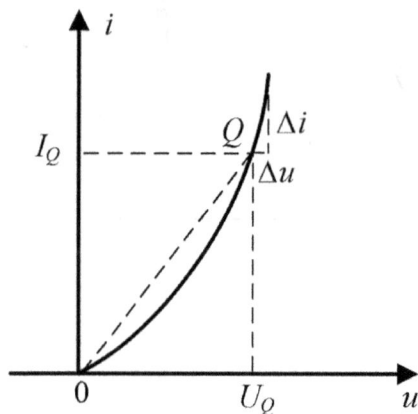

Fig. 1.38: Diagram of AC conductance.

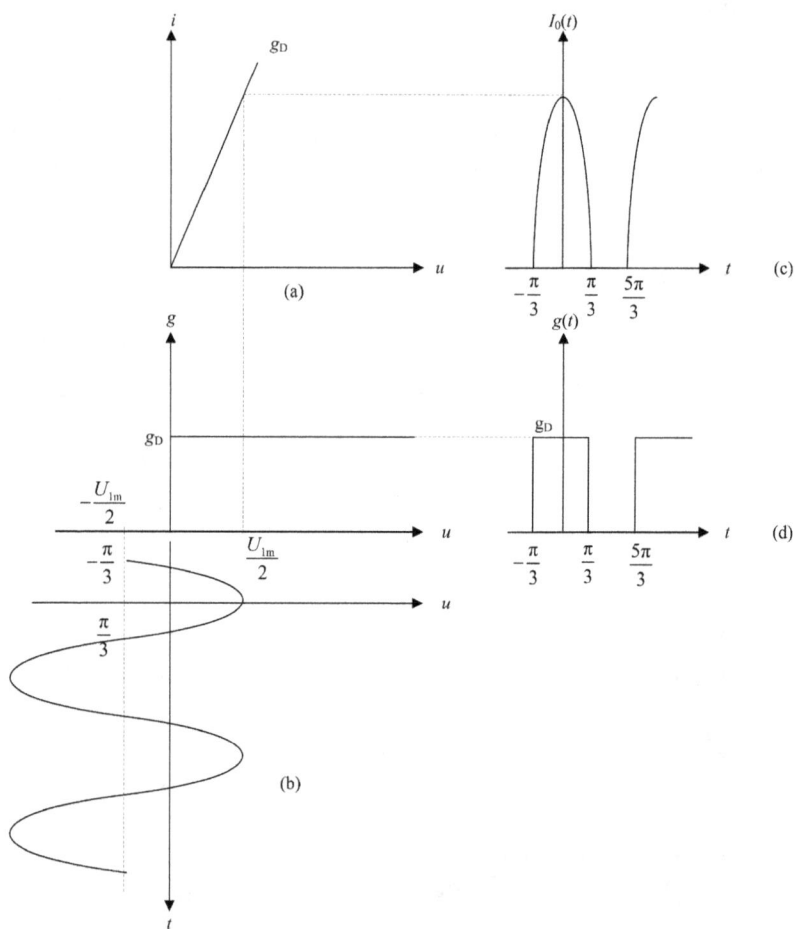

Fig. 1.39: Waveforms of $I_0(t)$ and $g(t)$ (a) nonlinear device i-u characteristic (b) input voltage with negative bias (c) output current waveforms (d) ON-OFF characteristic of the nonlinear device.

(1) DC conductance is also called the static conductance, which is defined as the slope between any point on the i–u curve and the origin, as shown in Fig. 1.37. DC conductance is frequently used in static analysis, denoted by g_0,

$$g_0 = \frac{I_Q}{U_Q} \qquad (1.87)$$

g_0 is a nonlinear function of I_Q or U_Q. For linear devices, g_0 value shall stay constant in the interested region.

(2) AC conductance is also called incremental conductance or differential conductance. It is defined as the tangential slope at any point on the i–u characteristic curve or the ratio of the incremental current to the incremental voltage at that specific point, as shown in Fig. 1.38. AC conductance is mainly used in small signal analysis, expressed as g_d:

$$g_d = \lim_{\Delta u \to \infty} \frac{\Delta i}{\Delta u} = \frac{di}{du}\bigg|_Q \qquad (1.88)$$

g_d is also a nonlinear function of U_Q or I_Q. AC conductance is used to analyze the current response to a small voltage. g_d can be considered linear in a very limited range. That is to say, a nonlinear resistor can be considered approximately as linear when the input voltage is small.

(3) Average conductance refers to the transient value when a large alternating signal is superimposed over a static voltage onto the nonlinear device. The transient slopes are different for the nonlinear resistor, which is mainly used in the large signal analysis, expressed as g_{av}:

$$g_{av} = \frac{I_{1m}}{U_m} \qquad (1.89)$$

where I_{1m} is the fundamental component of current, U_m is the amplitude of the applied voltage. g_{av} is not only related to the static point Q, but also to the magnitude of the signal source.

(4) To simplify analysis, supposing the device is linear and its conductance is g_D, the waveforms of output current $I_0(t)$ and conductance $g(t)$ from sinusoidal input voltage $u(t)$ is shown in Fig. 1.39. When the transistor is on, $g(t)$ is equal to g_D, and g is equal to zero as the transistor cut-off. Therefore, the waveform of $g(t)$ is like that of an electronic switch.

1.9.3 Frequency conversion by nonlinear devices

All nonlinear devices are capable of frequency conversion. Despite the detailed mathematical expressions, the nonlinear output in principle should provide multiple new frequency components given a sinusoidal (single frequency) input. Numerous functional circuits can be built based on this feature. For example, multiplication is a key operation of signals in communication electronic systems. Devices with good square law or multiplication property are extremely significant in applications. Here we assume a special case, where the i–u characteristic function of a nonlinear device obeys the perfect square law and the output current is only related to the second order of the input voltage:

$$i = au^2 \tag{1.90}$$

If two input signals u_1 and u_2 simultaneously enter this nonlinear device, the output current is then

$$i = a(u_1 + u_2)^2 = au_1^2 + 2au_1u_2 + au_2^2 \tag{1.91}$$

Assume $u_1 = U_{1m} \cos w_1 t$, $u_2 = U_{2m} \cos w_2 t$,

$$
\begin{aligned}
i &= aU_{1m}^2 \cos^2 w_1 t + 2aU_{1m}U_{2m} \cos w_1 t \cos w_2 t + aU_{2m}^2 \cos^2 w_2 t \\[4pt]
&= \frac{aU_{1m}^2}{2}(1 + \cos 2w_1 t) + \frac{aU_{2m}^2}{2}(1 + \cos 2w_2 t) \\[4pt]
&\quad + aU_{1m}U_{2m}[\cos(w_1 - w_2)t + \cos(w_1 + w_2)t] \\[4pt]
&= \frac{aU_{1m}^2}{2} + \frac{aU_{2m}^2}{2} + aU_{1m}U_{2m}\cos(w_1 + w_2)t \\[4pt]
&\quad + aU_{1m}U_{2m}\cos(w_1 - w_2)t + \frac{aU_{1m}^2}{2}\cos 2w_1 t + \frac{aU_{2m}^2}{2}\cos 2w_2 t
\end{aligned}
\tag{1.92}
$$

From eq. (1.92) it is observed that besides the first and second terms (DC components), several new frequency components are produced such as $(w_1 - w_2)$, $(w_1 + w_2)$, $2w_1$ and $2w_2$, during which frequency conversion is realized. And it is not hard to speculate that different i–u characteristic (or mathematical expression) will generate different frequency components as output. In order to understand frequency conversion results for more generalized cases, instead of a special square-law case, we will talk about several necessary mathematical techniques as follows.

1.9.3.1 Power series analysis

Let the i–u characteristic function of a nonlinear device be $i = f(u)$, where input $u = U_Q + u_1 + u_2$ (U_Q is the static operating point voltage while u_1 and u_2 are two AC signals). The characteristic function f can be any form which can be represented by Taylor series expansion at the proximity of U_Q

$$i = a_0 + a_1(u_1 + u_2) + a_2(u_1 + u_2)^2 + \cdots + a_n(u_1 + u_2)^n \tag{1.93}$$

Coefficient a_n of each term is given by

$$a_n = \frac{1}{n!}\frac{d^n f(u)}{du^n}\bigg|_{u=U_Q} = \frac{f^n(U_Q)}{n!} \tag{1.94}$$

Because

$$(u_1 + u_2)^n = \sum_{m=0}^{n}\frac{n!}{m!(n-m)!}u_1^{n-m}u_2^m \tag{1.95}$$

Equation (1.93) can be rewritten as

$$i = \sum_{n=0}^{\infty}\sum_{m=0}^{n}\frac{n!}{m!(n-m)!}a_n u_1^{n-m}u_2^m \tag{1.96}$$

The resulting AC current is the sum of numerous terms of $u_1^{n-m}u_2^m$. When $u_1 = U_{1m}\cos \omega_1 t$ and $u_2 = U_{2m}\cos \omega_2 t$, by setting $p = n-m$, $q = m$, the possible frequency combinations of eq. (1.96) are $|\pm p\omega_1 \pm q\omega_2|$ from the product of u_1^p and u_2^q.

Back to the multiplication operation we discussed in the first place, among all the frequency components, only when $m = 1$ and $n = 2$ we can obtain the expected result of $u_1 u_2$ multiplication with the frequency of $|\pm \omega_1 \pm \omega_2|$. Besides $|\pm \omega_1 \pm \omega_2|$ terms, other frequency components are unwanted signals and are regarded as noise. In order to eliminate these terms, the most suitable nonlinear device and the optimal operational point should be chosen to get the desired multiplication results maximized. In engineering applications, the other frequency terms can be suppressed by the balanced circuit, ring circuit and so on.

1.9.3.2 Linear time-varying analysis

A nonlinear function is given as $i = f(u)$, where $u = U_Q + u_1 + u_2$. We assume u_2 is a small signal so its second and higher order terms can be ignored. The Taylor expansion of u_2 at around $(U_Q + u_1)$ is

$$i = f(U_Q + u_1) + f'(U_Q + u_1)u_2 \tag{1.97}$$

The first term $f(U_Q + u_1)$ is a nonlinear function of u_1 and independent of u_2, which is called the static time-varying current and represented as $I_0(t)$. The second term coefficient $f'(U_Q + u_1)$ is the time-varying gain conductance. This conductance is also a nonlinear function of u_1 represented as $g(t)$. Equation (1.97) can be expressed as follows:

$$i = I_0(t) + g(t)u_2 \tag{1.98}$$

As observed from eq. (1.98), i and u_2 is approximately linear given that u_2 is small. However, the two coefficients $I_0(t)$ and $g(t)$ are time varying. The circuit is, therefore, called a linear time-varying circuit. Let $u_1 = U_{1m}\cos \omega_1 t$, then $I_0(t)$ and $g(t)$ can be expanded by Fourier series:

$$I_0(t) = I_{00} + I_{01}\cos \omega_1 t + I_{02}\cos 2\omega_1 t + \cdots + I_{0n}\cos n\omega_1 t \tag{1.99}$$

$$g(t) = g_0 + g_1 \cos w_1 t + g_2 \cos 2w_1 t + \cdots + g_n \cos nw_1 t \qquad (1.100)$$

The Fourier coefficients are calculated as follows:

$$
\begin{cases}
g_0 = \frac{1}{2\pi} \int_{-\pi}^{\pi} g(t) dw_1 t \\
g_1 = \frac{1}{\pi} \int_{-\pi}^{\pi} g(t) \cos w_1 t dw_1 t \\
\quad \vdots \\
g_n = \frac{1}{\pi} \int_{-\pi}^{\pi} g(t) \cos nw_1 t dnw_1 t
\end{cases}
\qquad (1.101)
$$

Let $u_2 = U_{2m} \cos w_2 t$ and plug u_2, $I_0(t)$ in eq. (1.99) and $g(t)$ in eq. (1.100) into eq. (1.98). The possible output frequency combinations can be found as

$$w = |\pm p w_1 \pm w_2| \qquad (1.102)$$

When $p = 1$, the direct multiplication of two input signals is realized. Compared to previous power series analysis, high order terms of w_2 in linear time-varying state as shown in eq. (1.102) are removed.

An example is provided here on how to apply the linear time-varying analysis method. The i–u characteristic of a nonlinear device is known as

$$
i = \begin{cases}
g_D u, & u \geq 0 \\
0, & u \leq 0
\end{cases}
\qquad (1.103)
$$

Input voltage $u = U_Q + U_{1m} \cos w_1 t + U_{2m} \cos w_2 t$ where $U_Q = -\frac{U_{1m}}{2}$ and U_{2m} is small enough to satisfy the linear time-varying operational premise. The waveform of $I_0(t)$ and $g(t)$ can be figured out by the graphic method.

According to eq. (1.103), the i–u characteristic curve is plotted in Fig. 1.39(a) with the slope being g_D. Slope g_D is a constant with respect to u as shown in Fig. 1.39(b). Since U_{2m} is so small and can be neglected, U_Q and $U_{1m} \cos w_1 t$ dominate the input u. Only when $u \geq 0$, current appears in the circuit. The corresponding waveforms of $I_0(t)$ and $g(t)$ are obtained in Fig. 1.39(c) and Fig. 1.39(d).

According to eqs. (1.100) and (1.101), coefficients of $g(t)$ are

$$g_0 = \frac{1}{2\pi} \int_{-\frac{\pi}{3}}^{\frac{\pi}{3}} g_D dw_1 t = \frac{g_D}{3} \qquad (1.104)$$

$$g_n = \frac{1}{\pi} \int_{-\frac{\pi}{3}}^{\frac{\pi}{3}} g_D \cos nw_1 t dnw_1 t = \frac{2g_D}{n\pi} \sin \frac{n\pi}{3} \qquad (1.105)$$

Therefore,

$$g(t) = \frac{g_D}{3} + \frac{2g_D}{\pi} \sum_{n=1}^{\infty} \frac{1}{n} \sin \frac{n\pi}{3} \qquad (1.106)$$

The expression of $I_0(t)$ can be calculated in the same way. Current I is obtained.

1.9.3.3 Exponential function analysis method

Assume the nonlinear device is a diode with the i–u characteristic function of

$$i = I_S \left(e^{\frac{qu}{kT}} - 1 \right) \bullet I_S e^{\frac{qu}{kT}} \tag{1.107}$$

where $u = U_Q + u_1 + u_2$. If diode works in the linear time-varying state, the current can be expressed as $i = I_0(t) + g(t)u_2$. Here,

$$I_0(t) = I_S e^{\frac{q(U_Q + u_1)}{kT}} = I_S e^{\frac{qU_Q}{kT}} e^{\frac{u_1}{kT}} = I_Q e^{\frac{u_1}{kT}} \tag{1.108}$$

Let $u_1 = U_{1m} \cos \omega_1 t$ and $I_0(t)$ is

$$I_0(t) = I_Q e^{\frac{U_{1m} \cos \omega_1 t}{kT}} \tag{1.109}$$

where $I_Q = I_S e^{\frac{qU_Q}{kT}}$.

$$g(t) = \left. \frac{\partial i}{\partial u} \right|_{u = U_Q + u_1} = \frac{qI_S}{kT} e^{\frac{q(U_Q + u_1)}{kT}} = \frac{qI_S}{kT} e^{\frac{qU_Q}{kT}} e^{\frac{qu_1}{kT}} = g_Q e^{\frac{qu_1}{kT}} \tag{1.110}$$

where $g_Q = \frac{qI_S}{kT} e^{\frac{qU_Q}{kT}}$.

The current i is then

$$i = I_0(t) + g(t)u_2 = I_Q e^{\frac{qU_{1m} \cos \omega_1 t}{kT}} + g_Q e^{\frac{qU_{1m} \cos \omega_1 t}{kT}} \tag{1.111}$$

The Fourier series expansion of $e^{\frac{qU_{1m} \cos \omega_1 t}{kT}}$ is

$$e^{\frac{qU_{1m} \cos \omega_1 t}{kT}} = a_0 \left(\frac{qU_{1m}}{kT} \right) + 2 \sum_{n=1}^{\infty} a_n \left(\frac{qU_{1m}}{kT} \right) \cos n\omega_1 t \tag{1.112}$$

Where, $a_n \left(\frac{qU_{1m}}{kT} \right)$ is the n-order Bessel function. Substitute eq. (1.112) in eq. (1.111) to get

$$i = (I_Q + g_Q u_2) \left[a_0 \left(\frac{qU_{1m}}{kT} \right) + 2 \sum_{n=1}^{\infty} a_n \left(\frac{qU_{1m}}{kT} \right) \cos n\omega_1 t \right] \tag{1.113}$$

The term of $2g_Q a_1 \left(\frac{qU_{1m}}{kT} \right) \cos \omega_1 t \cdot u_2$ reflects the effect resulting from the multiplication operation by the nonlinear circuit.

1.9.3.4 Switching function analysis method

Switching function analysis is the special case of linear time-varying analysis, which requires one input signal being small enough and the other being large enough. When the amplitude of the applied excitation signal is large enough, the dynamic range of the nonlinear device varies from cut-off to saturation. The i–u characteristic of the device can then be approximated by the piecewise lines, which is a switching behavior essentially. For example, in Fig. 1.40, u_1 and u_2 are two input signals where

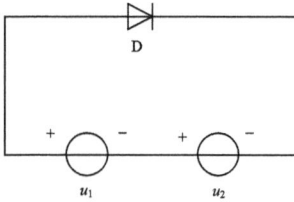

Fig. 1.40: Switching circuit of diode.

u_1 is large and u_2 is tiny. The diode then works in the alternating on and off switching state, which is controlled mainly by the large signal u_1. The on-state resistance (or the forward-conducting resistance) of the diode is R_D.

The graphic analysis procedure is shown in Fig. 1.41, which is similar to Fig. 1.39. The transistor is on for half an input signal period. When the input signal u(t) is sinusoidal, g(t) is a switch waveform with an amplitude of gD.

Here, we need to introduce the switching function $K_1(w_1t)$, which is an unidirectional periodic square wave with the normalized amplitude of unity (Fig. 1.42). $K_1(w_1t)$ is controlled by u_1, and the mathematical form of $K_1(w_1\,t)$ can be represented by the Fourier series expansion as

$$K_1(w_1t) = \frac{1}{2} + \frac{2}{\pi}\cos w_1t - \frac{2}{3\pi}\cos 3w_1t + \frac{2}{5\pi}\cos 5w_1t - \cdots$$

$$= \frac{1}{2} + \sum_{n=1}^{\infty}(-1)^{n-1}\frac{2}{(2n-1)\pi}\cos(2n-1)w_1t \tag{1.114}$$

$K_1(w_1t)$ has the same period with g(t) and $I_0(t)$, so the solution of the diode switching function in Fig. 1.40 is much simpler expressed by $K_1(w_1t)$:

$$g(t) = g_D K_1(w_1t) \tag{1.115}$$

$$I_0(t) = g_D K_1(w_1t)u_1 \tag{1.116}$$

Therefore

$$i = I_0(t) + g(t)u_2 = g_D(u_1 + u_2)K_1(w_1t) \tag{1.117}$$

When $u_1 = U_{1m}\cos w_1t$ and $u_2 = U_{2m}\cos w_2t$, according to eq. (1.117), the possible combinations of the frequency components are

$$w = |\pm(2n-1)w_1 \pm w_2| \tag{1.118}$$

Compared with the linear time-varying analysis, even-order terms of w_1 are gone and the useless terms are reduced by half.

In addition to the unidirectional switching function, there is also the bidirectional switching function $K_2(w_1t)$, which is also a periodic square wave with amplitude ±1 (Fig. 1.43). Suppose $K_2(w_1t)$ is also controlled by u_1.

Again, by Fourier series expansion $K_2(w_1t)$ can be written as

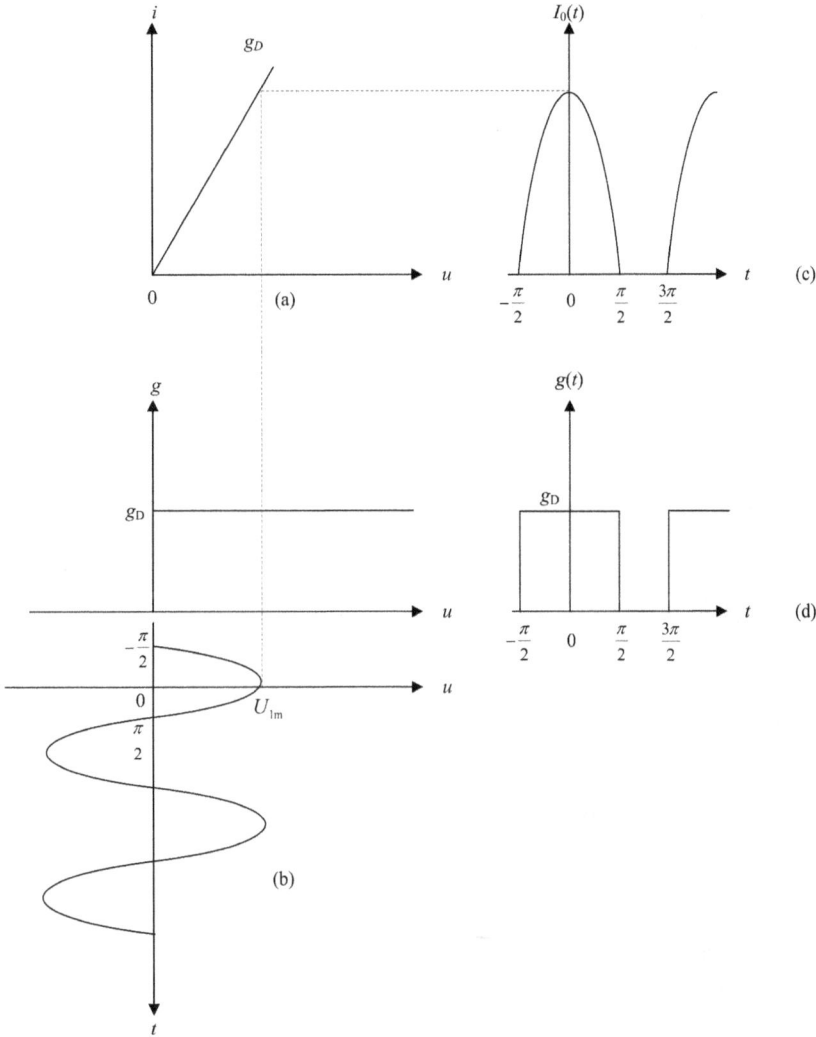

Fig. 1.41: Switching function analysis method. (a) ideal switching i-u characteristic, (b) input voltage waveform, (c) output current pulses, (d) switching pulses.

$$K_2(\omega_1 t) = \frac{4}{\pi} \cos \omega_1 t - \frac{4}{3\pi} \cos 3\omega_1 t + \frac{4}{5\pi} \cos 5\omega_1 t - \cdots$$

$$= \sum_{n=1}^{\infty} (-1)^{n-1} \frac{4}{(2n-1)\pi} \cos(2n-1)\omega_1 t \qquad (1.119)$$

The switching functions are known to follow the rules of

$$K_1(\omega_1 t) + K_1(\omega_1 t - \pi) = 1 \qquad (1.120)$$

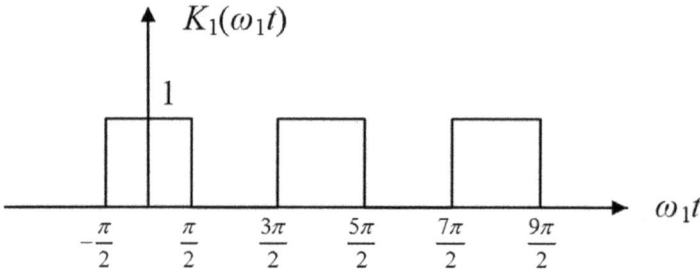

Fig. 1.42: Wave of unidirectional switching function.

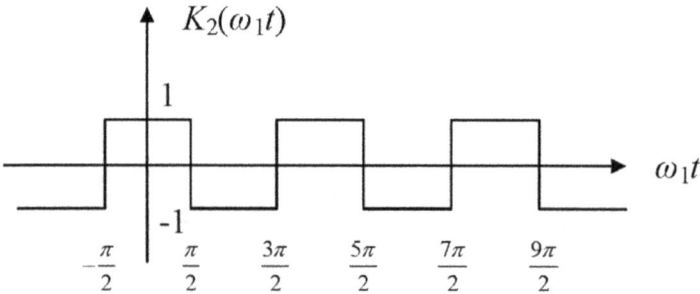

Fig. 1.43: Wave of bidirectional switching function.

$$K_1(\omega_1 t) - K_1(\omega_1 t - \pi) = K_2(\omega_1 t) \tag{1.121}$$

1.9.3.5 Hyperbolic function analysis

From the previous analysis, we know that useless frequency components are usually inevitable when signal multiplication is realized by frequency conversion. Analog multipliers can provide multiplication with nearly only two frequency components. Assume u_1 and u_2 are the input signals and the ideal output characteristic of the multiplier is

$$u_o = K u_1 u_2 \tag{1.122}$$

where K is the gain factor of the analog multiplier with the unit of V^{-1}. The symbol for analog multiplier is shown in Fig. 1.44.

The most popular analog multiplier used in high-frequency circuits is the four-quadrant variable transconductance analog multiplier. Its basic internal circuit structure is the differential amplifier circuit shown in Fig. 1.45.

In Fig. 1.45, the input signal $u_i = (u_{BE1} - u_{BE2})$ provides the base voltage for transistor T_1 and T_2. The output is obtained from the current difference at collectors. The current at both emitters can be written as

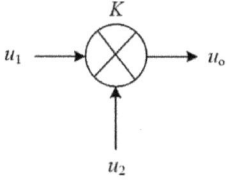

Fig. 1.44: Analog multiplier symbol.

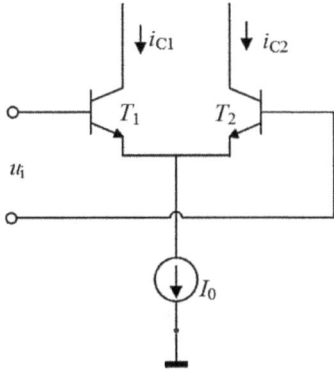

Fig. 1.45: Differential pair principle circuit.

$$i_{E1} = I_{S1}\left(e^{\frac{u_{BE1}q}{kT}} - 1\right) \approx I_{S1}e^{\frac{u_{BE1}q}{kT}} \tag{1.123}$$

$$i_{E2} = I_{S2}\left(e^{\frac{u_{BE2}q}{kT}} - 1\right) \approx I_{S2}e^{\frac{u_{BE2}q}{kT}} \tag{1.124}$$

Where, k is the Boltzmann's constant, q is the electronic charge, T is the absolute temperature, and I_S is the saturation current on the collector. When the two transistors T_1 and T_2 are completely symmetrical, $I_{S1} = I_{S2}$. The output current of differential pair is then

$$i = i_{C1} - i_{C2} \approx i_{E1} - i_{E2} = I_0 \tanh \frac{(u_{BE1} - u_{BE2})}{2kT} \tag{1.125}$$

If $u_i = (u_{BE1} - u_{BE2}) = U_m \cos \omega t$ is a small signal (which usually indicates $U_m < 26\text{mV}$), eq. (1.125) can be simplified as

$$i \approx (u_{BE1} - u_{BE2}) \frac{I_0 q}{2kT} \tag{1.126}$$

The current i is approximately linearly related to the input signal $u_i = (u_{BE1} - u_{BE2})$.
If the input is instead a large signal, that is, $U_m \geq 260\text{mV}$, the output current is

$$i = I_0 \tanh\left(\frac{qU_m}{2kT} \cos \omega t\right) = I_0 \tanh\left(\frac{x}{2} \cos \omega t\right) \tag{1.127}$$

The Fourier series expansion of (1.127) is

$$i = I_{1m}\cos\omega t + I_{3m}\cos 3\omega t + I_{5m}\cos 5\omega t + \cdots \tag{1.128}$$

This formula highly resembles a bidirectional switching function and the coefficients I_{nm} are listed in Table 1.4.

Table 1.4: Relationship between I_{nm}/I_0 and x.

x	I_{1m}/I_0	I_{3m}/I_0	I_{5m}/I_0
0.0	0.0000	0.0000	0.0000
0.5	0.1231	—	—
1.0	0.2356	−0.0046	—
1.5	0.3305	−0.0136	—
2.0	0.4058	−0.0271	—
2.5	0.4631	−0.0435	0.0026
3.0	0.5054	−0.0611	0.0097
4.0	0.5586	—	—
5.0	0.5877	−0.1214	0.0355
7.0	0.6112	−0.1571	0.0575
10.0	0.6257	−0.1827	0.0831
∞	0.6366	−0.2122	0.1273

Therefore, it can be concluded that when the input signal is small enough, the differential pair operates in the linear amplification state; when the input is large, the differential pair is close to a limiter circuit. Regardless of the operation state, if current I_0 is a controlled source such as the expression shown in eq. (1.129) by a voltage u_r, linear gain control can be anticipated:

$$I_0 = A + Bu_r \tag{1.129}$$

1.10 Summary

This chapter serves as an introduction to basic concepts as well as a connection or transition between analog electronic circuits to more complicated higher frequency circuits for communication electronic circuits. Main contents cover the composition of communication electronic circuit system, the operational process of receiver and transmitter, operational principle of LC resonant circuit including resonance condition, resonant characteristics, resonant curves and circuit applications, and characteristics of nonlinear devices. The following chapters are laid out to discuss key functional modules in a communication system one by one so that the readers can get comprehensive understandings.

Problems

1.1 What are the components of the wireless communication system? Write the functions of each part.

1.2 What is modulation? Why to carry on modulation?

1.3 Known in the parallel resonant circuit $f_0 = 5$ MHz, $C = 50$ nF, resonant resistor $R_0 = 100$ Ω. Calculate Q_0 and L. If the current amplitude of signal source is $I_{sm} = 0.1$ mA, calculate U_0 and current values of I_{C0} and I_{L0}.

1.4 What are the functions of frequency-selective networks in the communication system?

1.5 Series-resonant circuit is shown as Ex.Fig. 1.1, frequency of the signal source is 1 MHz, voltage amplitude $U_{sm} = 0.1$ V, $C = 100$ pF. During a short circuit at A–B, the circuit occurs resonance, and the voltage on capacitor C is 10 V. If a capacitive impedance Z_X is connected between A-B when $C = 200$ pF, circuit occurs resonance again, and at this time the voltage on capacitor is 2.5 V. Calculate L, Q_0, Q_L, $B_{0.7}$ and Z_X.

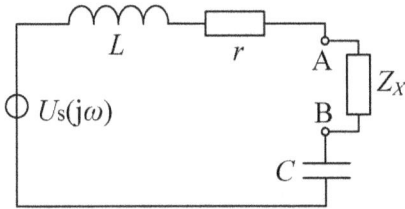 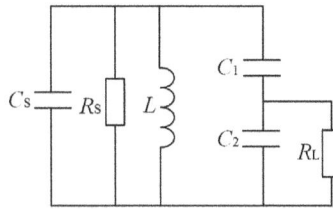

Ex.Fig. 1.1   **Ex.Fig. 1.2**

1.6 In Ex.Fig. 1.2, $L = 0.8$ µH, $C_1 = 20$ pF, $C_S = 10$ pF, $R_S = 20$ kΩ, $R_L = 5$ kΩ, $Q_0 = 100$. If R_L and R_S achieves impedance matching, what is the value of C_2? Calculate the resonant frequency f_0, resonant resistance R_0, Q_L and $B_{0.7}$.

1.7 We know that the resonant frequency $f_0 = 10.7$ MHz, $C_1 = C_2 = 15$ pF, $Q_0 = 100$, $R_L = 100$ kΩ, $R_{in} = 4$ kΩ, $C_{in} = 1$ pF, as shown in Ex.Fig. 1.3. If R_L and R_{in} are impedance matched, calculate N_1/N_2, L, resonant resistanceR_0, Q_L and $B_{0.7}$.

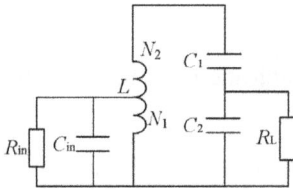

Ex.Fig. 1.3

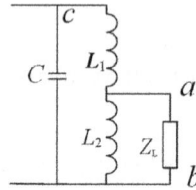

Ex.Fig. 1.4

1.8 Demonstration: in Ex.Fig. 1.4, $\dfrac{|Z_{ab}|}{|Z_{bc}|} = \dfrac{L_2^2}{(L_1+L_2)^2} = n^2$.

1.9* A T-type matching network is shown in Ex.Fig. 1.5, $R_L = 100\ \Omega$, $Q_1 = 3(Q_1 = \frac{X_{L1}}{R_L})$, $f = 60$ MHz, $R_1 = 200\ \Omega$. Calculate the values of C_1, C_2 and the inductor.

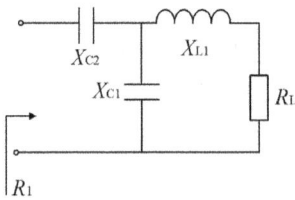

Ex.Fig. 1.5

Chapter 2
Radio frequency amplifiers

2.1 Introduction to radio frequency amplifiers

When small radio frequency (RF) signals arrive at the receiver antenna, they are fed into the input terminal of the RF weak signals amplifier through a passive matching network. Matching network can achieve maximum power transmission by equalizing the amplifier input impedance with the antenna impedance. Afterward, the RF amplifier's function is to increase the received signal power and get it ready for the follow-up process.

A large signal power amplifier (PA) is an important module, playing an essential action in the realization of many RF telecommunication systems. The role of an RFPA is to raise the input signal power level within a given frequency range up to a certain power level at its output terminal [5]. Comparison with low-level (i.e., linear) amplifiers is usually specified as small-signal gain and noise figure. However, the PA's performance metrics are the absolute output power level and power gain. A PA is therefore considered as a nonlinear system module. Its large-signal working conditions often result in adverse effects at its output signal, bringing about a distorted duplicate of the input signal.

In the first part of this chapter, we introduce small signal models of transistor and small signal tuned RF amplifier. In the second part of this chapter, we describe RF power amplifiers (RFPA). In this textbook, unless specifically stated, we note tuned amplifiers or resonate amplifiers as RF amplifiers.

2.2 Small signal tuned amplifier

A tuned amplifier is a bandpass amplifier. It has both frequency selection and signal amplifier functions. It can be worked both in low-level circuits and large-signal amplifiers, similar to the scenario for low-pass amplifiers. For a low-level amplifier, the desired general gain is a function of frequency. As a result, the response curve of a small signal amplifier is output magnitude versus frequency, as shown in Fig. 2.1.

It is the center frequency f_0 where the peak–magnitude response occurs. Commonly, the two frequencies where the magnitude decreases from the peak value to a given quantity are called the passband edges, and the interval between these two frequencies is the bandwidth of the passband. Usually, the band edges are defined as the −3 dB positions and the bandwidth is referred to as the −3 dB bandwidth.

https://doi.org/10.1515/9783110593822-002

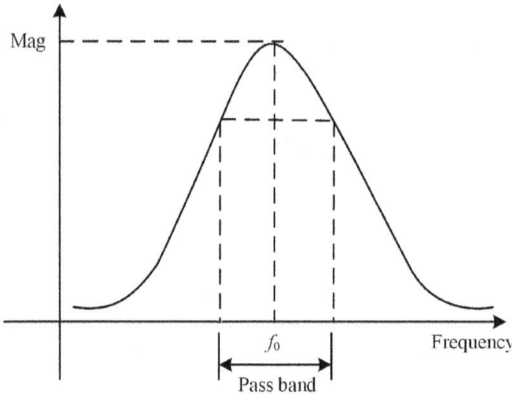

Fig. 2.1: Frequency response of the general gain.

2.2.1 High-frequency model of transistor

The primary function of a "model" is to predict the behavior of a device in a particu-
lar operating region. The behavior of the bipolar junction transistor (BJT) in the
high-frequency small-signal domain is quite different from its low-frequency cases.
The small-signal alternating current (AC) response can be described by two common models: the Y-parameters model and the hybrid π model [3]. The models are
mathematically equivalent that allows us to predict circuit performance.

2.2.1.1 Y-parameter BJT equivalent circuit

Figure 2.2 shows the BJT and its Y-parameters equivalent circuit. Let us assume that
the input voltage is u_1, the input current is i_1, the output voltage is u_2 and the output
current is i_2. The input current source $y_r u_2$ is a function of the output collector–emitter
voltage and the output current source $y_f u_1$ is a function of the input base voltage. The
input conductance is y_i and the output conductance is y_o. Parameters of y_i, y_r, y_f and
y_o are Y parameters of the BJT.

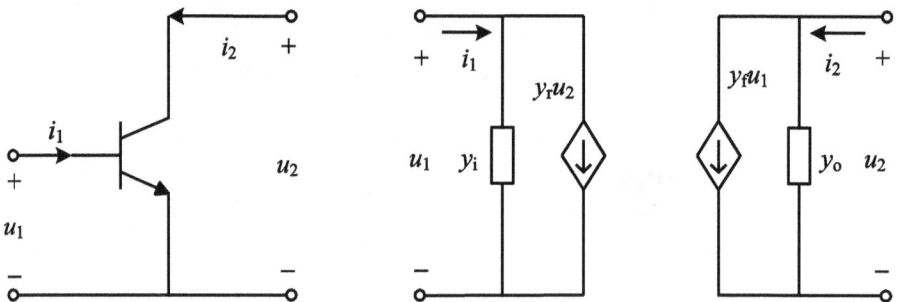

Fig. 2.2: BJT and its y parameter circuit.

To describe the device behavior in an active mode, we can write

$$i_1 = y_{11}u_1 + y_{12}u_2$$
$$i_2 = y_{21}u_1 + y_{22}u_2$$

$$(2.1)$$

In eq. (2.1) the coefficients or Y parameters are defined as

$y_{11} = y_i = \dfrac{i_1}{xu_1}\Big|_{u_2=0}$ is the input admittance when output is shorted,

$y_{12} = y_r = \dfrac{i_1}{u_2}\Big|_{u_1=0}$ is the inverse transfer admittance when input is shorted,

$y_{21} = y_f = \dfrac{i_2}{u_1}\Big|_{u_2=0}$ is the forward transfer admittance when output is shorted,

$y_{22} = y_o = \dfrac{i_2}{u_2}\Big|_{u_1=0}$ is the output admittance when input is shorted.

According to eq. (2.1), we can present the equivalent circuit in Fig. 2.2. For common-emitter amplifiers, $i_1 = i_b$, $u_1 = u_{be}$, $i_2 = i_c$, $u_2 = u_{ce}$, Y parameters are y_{ie}, y_{re}, y_{fe} and y_{oe}. For common-base amplifiers, $i_1 = i_e$, $u_1 = u_{eb}$, $i_2 = i_c$, $u_2 = u_{cb}$, Y parameters are y_{ib}, y_{rb}, y_{fb}, y_{ob}. For common-collector amplifiers, $i_1 = i_b$, $u_1 = u_{be}$, $i_2 = i_e$, $u_2 = u_{ec}$, Y parameters are y_{ic}, y_{rc}, y_{fc}, y_{oc}.

2.2.1.2 Hybrid π equivalent circuit of BJT

For high-frequency cases, the hybrid π equivalent circuit of transistor is shown in Fig. 2.3. When the transistor works at the amplifying state, the emitter junction is forward biased and the emitter junction resistance $r_{b'e}$ is small. From the H parameters model of transistor, we can learn

$$r_{b'e} = \frac{26\beta_0}{I_{EQ}}$$

$$(2.2)$$

where β_0 is the low-frequency current amplification factor of the common emitter transistor and I_{EQ} is the emitter quiescent current of the transistor. Base spreading

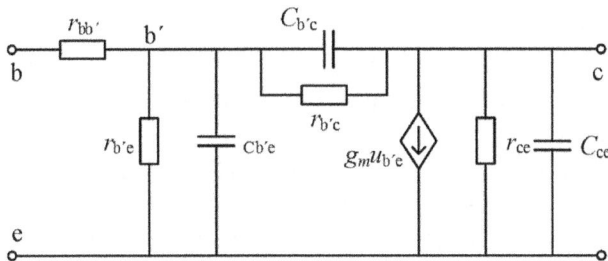

Fig. 2.3: High-frequency hybrid π equivalent circuit of BJT.

resistance and base lead resistance $r_{bb'}$ is generally dozens of ohms. The collector junction resistance is $r_{b'c}$ and the output collector emitter resistance is r_{ce}. The resistance $r_{b'c}$ is very high because of the reverse biasing of the collector junction. The emitter junction capacitance is $C_{b'e}$ and the collector junction capacitance is $C_{b'c}$, which is small comparing to $C_{b'e}$. The capacitance between the collector and the emitter is C_{ce}. The linear current source $g_m u_{b'e}$ is a function of base input voltage $u_{b'e}$. $g_m = I_{EQ}/V_T$ is the transconductance of the transistor, where V_T is the voltage temperature equivalence ($V_T \approx 26$ mV at 27 °C). The transconductance g_m reflects the amplify ability of the transistor [6].

In the working frequency range of transistor, the resistance $r_{b'c}$ is much higher than the AC resistance of $C_{b'c}$. So, in practice, we can use $C_{b'c}$ solely instead of the parallel circuit of $C_{b'c}$ and $r_{b'c}$.

In general, Y-parameter transistor model is applied when analyzing small signal resonant amplifiers. But Y parameters are not constant, which vary with frequency. Unfortunately, the exact measurement of Y parameters is extremely difficult in laboratory. Y-parameter models cannot explain explicitly the internal physical process of transistors. For these reasons, hybrid π equivalent circuit model is used frequently when analyzing the circuit's working principle. Hybrid π model employs lumped elements such as RC to analyze the circuits [7–8]. The physical process of transistor is clear. We can obtain Y parameters once we know the hybrid π equivalent circuit model parameters. The parameters can be connected by the following equations:

$$y_{ie} \approx \frac{g_{b'e} + j\omega C_{b'e}}{1 + r_{bb'}(g_{b'e} + j\omega C_{b'e})} \tag{2.3}$$

$$y_{re} \approx \frac{-j\omega C_{b'c}}{1 + r_{bb'}(g_{b'e} + j\omega C_{b'e})} \tag{2.4}$$

$$y_{fe} \approx \frac{g_m}{1 + r_{bb'}(g_{b'e} + j\omega C_{b'e})} \tag{2.5}$$

$$y_{oe} \approx \frac{j\omega C_{b'c} I_{bb'} g_m}{1 + r_{bb'}(g_{b'e} + j\omega C_{b'e})} + j\omega C_{b'c} \tag{2.6}$$

2.2.1.3 High-frequency parameters of the transistor

The BJT transistor has two PN junctions, the emitter junction and the collector junction. The movement of carriers in BJT transistor includes the following procedures. Electrons are released from the emitter and enter the base region with a small number of electrons combining with the holes at the base. Most of the electrons keep passing the base region and arrive at the collector. The electrons in the collector region enter the external circuitry under the applied voltage and form the collector current. When a transistor is operated at low frequencies, the effect of PN junction capacitance can be ignored; when at a higher frequency, the PN junction capacitance affects carrier movement significantly. The signal amplitude will reduce and the collector current amplification β will decrease as frequency increases. The cut-off frequency is defined as f_β when β value equals $1/\sqrt{2}\,\beta_0$. The relationship between β and β_0 is then

$$|\beta| = \frac{\beta_0}{\sqrt{1 + \left(\frac{f}{f_\beta}\right)^2}} \tag{2.7}$$

The characteristic frequency is called f_T at which β value reduces to 1. The relationship between f_T and f_β is

$$f_T = f_\beta \sqrt{\beta_0^2 - 1} \tag{2.8}$$

When $\beta_0 \gg 1$, $f_T \approx \beta_0 f_\beta$.

When $\beta_0 \gg 1$ and $f \gg f_\beta$, the relation between β and f is

$$|\beta| = \frac{\beta_0}{\sqrt{1 + \left(\frac{f}{f_\beta}\right)^2}} \approx \frac{\frac{f_T}{f_\beta}}{\sqrt{1 + \left(\frac{f}{f_\beta}\right)^2}} \approx \frac{\frac{f_T}{f_\beta}}{\frac{f}{f_\beta}} = \frac{f_T}{f} \tag{2.9}$$

If the frequency keeps rising to f_{max}, the power gain of the transistor is lowered to 1. f_{max} is called transistor's maximum frequency. The f_{max} is

$$f_{max} \approx \frac{1}{2\pi} \sqrt{\frac{g_m}{4r_{bb'} C_{b'e} C_{b'c}}} \tag{2.10}$$

where g_m is the transconductance of BJT, $r_{bb'}$ is the resistance of the base region, $C_{b'e}$ is the emitter junction capacitance and $C_{b'c}$ is the collector junction capacitance of BJT.

When designing a circuit, we can select a BJT whose f_T is three to five times of $f_{working}$, that is

$$f = \left(\frac{1}{3} - \frac{1}{5}\right) f_T \tag{2.11}$$

The frequency characteristic curve of a BJT transistor is shown as Fig. 2.4.

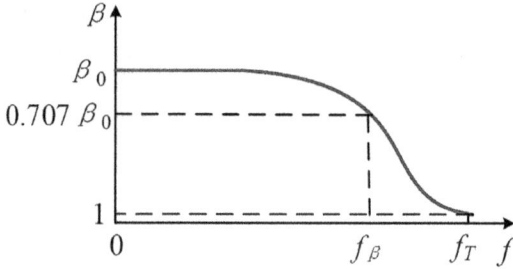

Fig. 2.4: Frequency characteristic of a BJT.

Example 2.1: A BJT's f_T is 500 MHz, $\beta_0 = 100$. Please calculate β values at $f = 2$ MHz, 20 MHz, and 200 MHz respectively.

Solution: Since $\beta_0 = 100 \gg 1$,

According to eq. (2.8), the cut-off frequency f_β is

$$f_\beta \approx \frac{f_T}{\beta_0} = \frac{500}{100} = 5 (\text{MHz})$$

(1) When $f = 2$ MHz, β value is

$$|\beta| = \frac{\beta_0}{\sqrt{1 + \left(\frac{f}{f_\beta}\right)^2}} = \frac{100}{\sqrt{1 + \left(\frac{2}{5}\right)^2}} \approx 92.6$$

$$\beta \approx \frac{f_T}{f} = \frac{500}{200} = 2.5$$

In fact, 2 MHz is low frequency compared to f_β. So, β value approximately equals to $\beta_0 = 100$. The relative deviation is 7.4%.

(2) When $f = 20$ MHz, β value is

$$|\beta| = \frac{\beta_0}{\sqrt{1 + \left(\frac{f}{f_\beta}\right)^2}} = \frac{100}{\sqrt{1 + \left(\frac{20}{5}\right)^2}} \approx 24.3$$

(3) When $f = 200$ MHz $\gg f_\beta$, β value is

$$\beta \approx \frac{f_T}{f} = \frac{500}{200} = 2.5$$

From the above calculations, we can see that β value decreases as the frequency increases.

2.2.2 Performance of small signal amplifier

For small signal amplifiers, the specifications concerned are as follows.

2.2.2.1 Gain

Gain is the ratio between the output voltage (or power) and the input voltage (or power) of the amplifier. The typical unit is dB. Gain can be written as

$$A_u = 20 \lg \frac{u_o}{u_i} \quad (\text{dB}) \tag{2.12}$$

$$A_P = 20 \lg \frac{P_o}{P_i} \quad (\text{dB}) \tag{2.13}$$

Large gain at center frequency or passband edge frequencies is highly welcomed since it avoids too many stages of amplifiers. Amplifier gain is determined by the transistor, bandwidth, matching network and working stability.

2.2.2.2 Bandwidth

Similar to a parallel resonant circuit, the illustration of passband of amplifiers is in Fig. 2.5. If we consider amplification, the band edges are defined in terms of where the $0.7A_0$ (A_0 refers to the largest magnitude) data points locate and the corresponding frequency separation is the bandwidth known as $BW_{0.7}$. In the same way, we define the other bandwidth $BW_{0.1}$ by the frequency separation between two $0.1A_0$ data

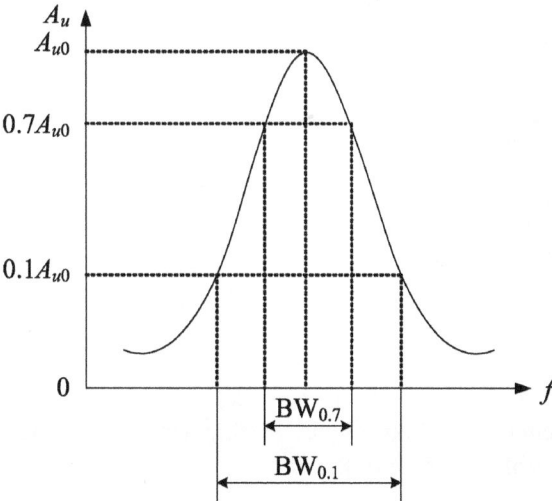

Fig. 2.5: Amplitude versus frequency curve of a resonant amplifier.

points. Obviously, $BW_{0.1}$ is wider than $BW_{0.7}$. Bandwidth reflects the selectivity of an amplifier. Bandwidth being smaller manifests the better selectivity.

If we normalize the magnification of the amplifier, we can write the normalized gain as

$$\frac{A_u}{A_0} = \frac{1}{\sqrt{1 + \left(\frac{2Q_L \Delta f_{0.7}}{f_0}\right)^2}} \tag{2.14}$$

where Q_L is the quality factor with load of the resonant circuit, $2\Delta f_{0.7}$ is the bandwidth $BW_{0.7}$ and f_0 is the center frequency of the amplifier. In eq. (2.14), when $\frac{A_u}{A_0} = \frac{1}{\sqrt{2}}$, we can get

$$2\Delta f_{0.7} = \frac{f_0}{Q_L} \tag{2.15}$$

2.2.2.3 Rectangle coefficient

As mentioned above, the selectivity of the amplifier can be described by the rectangle coefficient as

$$K_{r0.1} = \frac{2\Delta f_{0.1}}{2\Delta f_{0.7}} \tag{2.16}$$

Where, $2\Delta f_{0.1}$ is the bandwidth $BW_{0.1}$. It is not hard to observe that $K_{r0.1} > 1$.

When substituting $\frac{A_u}{A_0} = 0.1$ into eq. (2.14)

$$2\Delta f_{0.1} = \sqrt{10^2 - 1}\, \frac{f_0}{Q_L} \tag{2.17}$$

According to eq. (2.15), $K_{r0.1}$ can be obtained

$$K_{r0.1} = \frac{2\Delta f_{0.1}}{2\Delta f_{0.7}} = \sqrt{10^2 - 1} \approx 9.95 \tag{2.18}$$

These results show that the rectangle coefficient of a single resonant circuit amplifier is close to 10 suggesting poor selectivity.

2.2.3 Small signal tuned amplifier circuits

A single resonant amplifier is shown in Fig. 2.6. It is composed of a common emitter transistor and a shunt resonant circuit. R_1, R_2 and R_e are biasing resistors. C_e is the emitter shunt capacitor. Input signal u_i is fed into the base and output signal u_o is obtained at the collector.

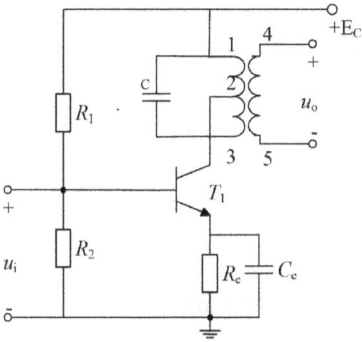

Fig. 2.6: A single resonant amplifier.

2.2.3.1 Equivalent circuit of the amplifier and simplified circuit

Figure 2.7 shows the equivalent circuit of a high-frequency small signal amplifier in Fig. 2.6. In Fig. 2.7, transistor T_1 is represented by its Y-parameter equivalent circuit, and signal source is substituted by I_s and Y_s. The transformer load y_{ie2} is the input admittance of the next amplifier.

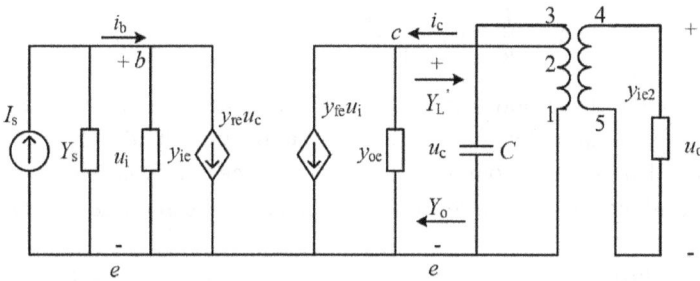

Fig. 2.7: High-frequency equivalent circuit of the single resonant amplifier.

We assume that when looking into the resonant circuit from between c and e the equivalent load admittance is Y_L'. Then, when Y_L' and I_s are connected with the transistor, we can write the follow relations according to the transistor internal characteristic.

$$i_b = y_{ie}u_i + y_{re}u_c \tag{2.19}$$

$$i_c = y_{fe}u_i + y_{oe}u_c \tag{2.20}$$

i_c is determined by the external load and has the form of

$$i_c = -Y_L'u_c \tag{2.21}$$

From eqs. (2.20) and (2.21), we can eliminate i_c and get

$$u_c = - \frac{y_{fe}}{y_{oe} + Y_L} u_i \tag{2.22}$$

From eqs. (2.19) and (2.22), u_c is replaced then we can get

$$i_b = y_{ie} u_i + y_{re} \left(- \frac{y_{fe}}{y_{oe} + Y_L'} u_i \right) \tag{2.23}$$

The input admittance of the amplifier can be expressed by Y parameters as

$$Y_i = \frac{i_b}{u_i} = y_{ie} - \frac{y_{fe} y_{re}}{y_{oe} + Y_L'} \tag{2.24}$$

Equation (2.24) shows that the second item always exists with a nonzero y_{re} ($y_{fe} \neq 0$). The input admittance Y_i of an amplifier does not only depend on the input admittance of the transistor y_{ie}, but also depends on the load Y_L' of the amplifier. That is to say, the input admittance Y_i varies when the load admittance Y_L' changes.

From eqs. (2.19), (2.20) and (2.21), the output admittance of the amplifier can be found too.

$$Y_o = \frac{i_c}{u_c} = y_{oe} - \frac{y_{fe} y_{re}}{y_{ie} + Y_s} \tag{2.25}$$

Equation (2.25) shows that the second item is nonzero as long as y_{re} is nonzero. The output admittance Y_o of an amplifier is not only related to the output admittance of the transistor y_{oe}, but also the input source admittance Y_s of the amplifier. That is to say, the input source admittance Y_s can result in the change of the amplifier output admittance Y_o.

To simplify the analysis of an amplifier, we assume $y_{re} = 0$. This means that the output terminal has no influence on the input terminal of the transistor. The equivalent circuit is only the output part shown in Fig. 2.8. We assume that the second stage amplifier using transistor T_2 with input admittance y_{ie2} is identical to the first

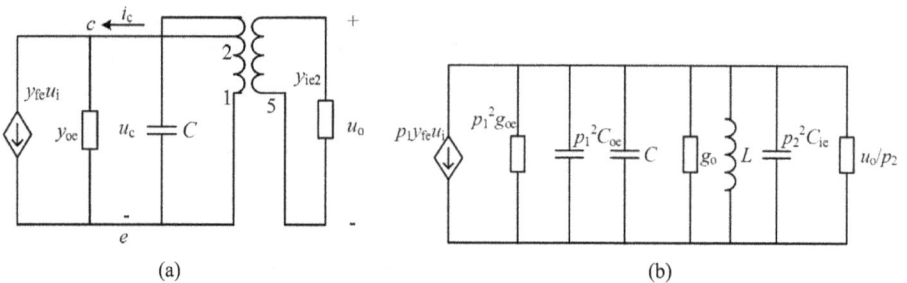

(a)

(b)

Fig. 2.8: Simplified equivalent circuit of the single resonant amplifier, (a) unidirectional device output equivalent circuit, (b) access factor expansion equivalent circuit.

transistor T_1 with input admittance y_{ie1}. When passing the two transistors collector current are equal, $y_{ie1} = y_{ie2} = y_{ie}$.

Assume the two transistors T_1 and T_2 are the same, the inductance of the coil inductor is L, and the unloaded quality factor at working frequency is Q_0. The resonant conductance can be obtained by

$$g_0 = \frac{1}{\omega_0 L Q_0} \tag{2.26}$$

The input and output admittance of a transistor can be regard as a resistor and a capacitor in parallel. The expression can be written as $y_{ie} = g_{ie} + j\omega C_{ie}$ and $y_{oe} = g_{oe} + j\omega C_{oe}$. The access coefficient of the transformer left coil is $p_1 = N_{12}/N_{13}$, and the access coefficient of the transformer right coil is $p_2 = N_{45}/N_{13}$. According to the simplified equivalent circuit in Fig. 2.7, we can analyze the technical specifications of the amplifier.

2.2.3.2 Amplifier technical specifications calculation

(1) Voltage gain A_u. According to Fig. 2.7 and voltage gain definition, the total admittance Y_Σ can be written as

$$Y_\Sigma = g_\Sigma + j\omega C_\Sigma + \frac{1}{j\omega L} \tag{2.27}$$

Where, $g_\Sigma = p_1^2 g_{oe} + g_0 + p_2^2 g_{ie}$ and $C_\Sigma = p_1^2 C_{oe} + C + p_2^2 C_{ie}$. The equivalent circuit contains a dependent current source $p_1 y_{fe} u_i$ with shunt admittance Y_Σ to provide the voltage of u_o/p_2. This can be expressed by

$$\frac{u_o}{p_2} = \frac{p_1 y_{fe} u_i}{Y_\Sigma} = -\frac{p_1 y_{fe} u_i}{g_\Sigma + j\omega C_\Sigma + \frac{1}{j\omega L}} \tag{2.28}$$

As a result

$$A_u = \frac{u_o}{u_i} = -\frac{p_1 p_2 y_{fe}}{g_\Sigma + j\omega C_\Sigma + \frac{1}{j\omega L}} \tag{2.29}$$

As for the amplifier, the resonant frequency is $\omega_0 = \frac{1}{\sqrt{LC_\Sigma}} \left(\omega_0 C_\Sigma = \frac{1}{\omega_0 L} \right)$. Then, we can give the gain at resonance as

$$A_{u0} = -\frac{p_1 p_2 y_{fe}}{g_\Sigma} \tag{2.30}$$

Where, the minus sign means that the output voltage is almost out of phase with the input voltage. Analyzing it further, we can find that y_{fe} is a complex number with the phase part of φ_{fe}. When frequency is very low, $\varphi_{fe} = 0$ suggesting that the output and input voltage is out of phase completely (phase difference = 180°). In other cases, output and input is not with exact 180° phase difference.

(2) Resonance curve. The resonance curve describes the relationship between the amplifier relative voltage gain and the input frequency. From eq. (2.29), we know that

$$A_u = -\frac{p_1 p_2 y_{fe}}{g_\Sigma \left[1 + \frac{1}{g_\Sigma}\left(j\omega C_\Sigma + \frac{1}{j\omega L}\right)\right]} = \frac{A_{u0}}{1 + j\frac{1}{\omega_0 L g_\Sigma}\left(\omega C_\Sigma \omega_0 L - \frac{\omega_0 L}{\omega L}\right)}$$

$$= \frac{A_{u0}}{1 + jQ_L\left(\frac{\omega}{\omega_0} - \frac{\omega_0}{\omega}\right)} \tag{2.31}$$

Therefore,

$$\frac{A_u}{A_{u0}} = \frac{1}{1 + jQ_L\left(\frac{\omega}{\omega_0} - \frac{\omega_0}{\omega}\right)} = \frac{1}{1 + jQ_L\left(\frac{f}{f_0} - \frac{f_0}{f}\right)}$$

When $f \approx f_0$, let $\Delta f = f - f_0$

$$\frac{A_u}{A_{u0}} = \frac{1}{1 + jQ_L\frac{2\Delta f}{f_0}} \tag{2.32}$$

where Δf is known as the detuning frequency.

We define $\xi = Q_L\frac{2\Delta f}{f_0}$, which is the generalized detuning factor. Substitute ξ into eq. (2.32) then the relative voltage gain is

$$\frac{A_u}{A_{u0}} = \frac{1}{1 + j\xi} \tag{2.33}$$

The magnitude or norm of the relative gain is

$$\frac{A_u}{A_{u0}} = \frac{1}{\sqrt{1 + \xi^2}} \tag{2.34}$$

(3) Bandwidth and rectangle coefficient. According to eq. (2.34), if we set $\xi = 1$, −3 dB bandwidth or $2\Delta f_{0.7}$ is obtained similarly as eq. (2.15). If $\xi = \sqrt{10^2 - 1}$, we can obtain $2\Delta f_{0.1}$ as eq. (2.17). After that the rectangle coefficient of the amplifier can be calculated as what eq. (2.18) defines. For a single resonant amplifier, the rectangle coefficient is approximate 9.95.

Example 2.2: The AC equivalent circuit of a small signal resonant amplifier is given in Fig. 2.9. The resonant frequency f_0 = 5 MHz, passband $B_{0.7}$ = 100 kHz, resonant voltage gain A_{u0} = 80. The Y parameters tested at the operating frequency are as follows,

y_{ie} = (1.8 + j0.6) mS, $y_{re} \approx 0$, y_{fe} = (15 − j8) mS, y_{oe} = (30 + j40) μS. $p_1 = N_{12}/N_{13}$ = 0.75, N_{45}/N_{13} = 0.8.

The quality factor of L without load Q_0 equals 100. Please find out the resonant tank parameters of L, C and the external resistance R.

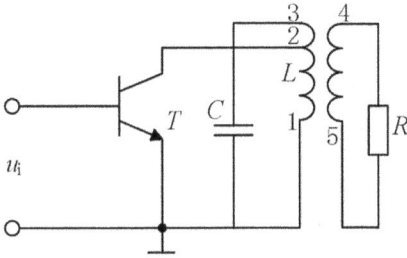

Fig. 2.9: AC equivalent circuit of small signal amplifier.

Solution: From Fig. 2.9, we can see that L is partially connected to the collector of the transistor. The access coefficients of L and transformer are $p_1 = N_{12}/N_{13} = 0.75$ and $p_2 = N_{45}/N_{13} = 0.8$, respectively.

From eq. (2.25), the voltage gain at resonant frequency is

$$\therefore A_{u0} = \frac{|p_1 p_2 y_{fe}|}{g_\Sigma} \Rightarrow g_\Sigma = \frac{|p_1 p_2 y_{fe}|}{A_{u0}} = \frac{0.75 \times 0.8 \times \sqrt{15^2 + 8^2}\,\text{mS}}{100} = 1.02 \times 10^{-4}\,\text{S}$$

The quality factor with load is

$$Q_L = \frac{f_0}{B_{0.7}} = \frac{5\text{MHz}}{100\,\text{kHz}} = 50$$

From the total conductance formula, inductance L satisfies

$$g_\Sigma = \frac{1}{Q_L \omega_0 L} \Rightarrow L = \frac{1}{Q_L \omega_0 g_\Sigma} = \frac{1}{50 \times 2\pi \times 5 \times 10^6 \times 1.02 \times 10^{-4}} \approx 6.24\,(\mu\text{H})$$

According to the LC parallel resonant circuit calculation formula, we can get C

$$C = \frac{1}{4\pi^2 f_0^2 L} = \frac{1}{4\pi^2 \times 25 \times 10^{12} \times 6.24 \times 10^{-6}} \approx 162.5\,(\text{pF})$$

The parallel-connected R satisfies

$$g_\Sigma = \frac{1}{Q_0 \omega_0 L} + \frac{1}{R} + g_{oe} \Rightarrow \frac{1}{R} = 1.02 \times 10^{-4} - \frac{1}{100 \times 2\pi \times 5 \times 10^6 \times 6.24 \times 10^{-6}} - 30 \times 10^{-6}$$

$$= 0.21 \times 10^{-4}\,(\text{s})$$

$$R = 47.6\text{k}\Omega$$

2.2.3.3 Multistage single resonant circuit amplifiers

A multistage single resonant amplifier is composed of several single resonant amplifiers. Higher gain and better performance can be achieved with such arrangement [9].

(1) Voltage gain. Suppose there are n-stage amplifiers with the gain of A_{u1}, A_{u2}, ... and A_{un}. The total gain of the amplifier should be the product of every single stage gain. It can be given as

$$A_n = A_{u1} \cdot A_{u2} \cdot \ \cdots \ \cdot A_{un} \tag{2.35}$$

If the unit of gain is in dB, then the total gain becomes the sum of all stages. That is

$$A_n(dB) = A_{u1}(dB) + A_{u2}(dB) + \cdots + A_{un}(dB) \tag{2.36}$$

(2) Resonance curve. If n stages of identical amplifiers are connected one after another, the output of $(m\text{-}1)$th becomes the input of the mth amplifier. Therefore, the total resonance curve equation is the product of every single resonance curve equation, which can be given by

$$\frac{A_{un}}{A_{u0}} = \frac{1}{\left[1 + \left(Q_L \frac{2\Delta f}{f_0}\right)^2\right]^{\frac{n}{2}}} \tag{2.37}$$

(3) Bandwidth. According to eq. (2.37), the total passband for n-stage amplifier is calculated as

$$(BW_{0.7})_n = (2\Delta f_{0.7})_n = \sqrt{2^{\frac{1}{n}} - 1}\,\frac{f_0}{Q_L} \tag{2.38}$$

When $n > 1, \sqrt{2^{1/n} - 1} < 1$. The total bandwidth of n-stage amplifiers is narrower than a single resonant amplifier.

(4) Rectangle coefficient. In eq. (2.37) we set $A_{un}/A_{u0} = 0.1$. Then the rectangle coefficient of the n-stage amplifier can be obtained as

$$(K_{r0.1})_n = \frac{\sqrt{100^{\frac{1}{n}} - 1}}{\sqrt{2^{\frac{1}{n}} - 1}} \tag{2.39}$$

Equation (2.39) shows that $K_{r0.1}$ becomes smaller as stage n gets larger.

2.2.4 Stability of small signal amplifiers

In real cases, the inverse transfer admittance y_{re} is not zero. There is an internal feedback when a transistor is linked to the circuit. This means that part of the output voltage of the amplifier will be fed back into the input terminal. This results in changes and instability of the transistor input current. In circuit systems, the feedback may lead to self-excited oscillations. A stability coefficient must be satisfied to ensure the amplifier to work in a stable condition. We encourage our readers to refer to other references for further discussions on this.

As mentioned earlier, y_{re} is not 0 and BJT is a bidirectional device. By removing internal feedback, or realizing unidirectional devices, the stability of amplifiers can be improved. There are two approaches to get a unidirectional device, one is neutralization method and another is mismatch method.

Neutralization method employs an additional external feedback loop circuit to compensate the internal feedback circuit of the transistor.

Mismatch method assists the circuit to meet the stability condition by decreasing the voltage gain of the amplifier. The most popular practice is a combination of two single-stage amplifiers, the common emitter followed by the common base. Such circuit is also known as the "cascode" amplifier [10].

2.2.5 Low noise variable gain broadband amplifiers

In recent years, large-scale integrated circuit has gained prominent development. The resonant amplifiers can be realized by integrated circuit broadband amplifier and lumped filters such as ceramic filter and surface acoustic wave filter.

Manufacturers have developed many types of wideband amplifiers. In this section, we will introduce a low noise 90 MHz variable gain amplifier AD603 made by the analog devices (AD) cooperation.

2.2.5.1 Features of AD603
AD603 is a linear-in-dB gain control, pin-programmable gain, bandwidth independent of variable gain, high gain accuracy RFIC. It can be utilized in RF amplifiers, intermediate frequency (IF) amplifiers, video gain control or signal measurement.
(1) −11 dB to +31 dB with 90 MHz bandwidth.
(2) 9–51 dB with 9 MHz bandwidth.
(3) Any medium range, such as −2 dB to +40 dB with 30 MHz bandwidth.
(4) 1.3 nV/√Hz input noise spectral density. ± 0.5 dB typical gain accuracy.
(5) Input resistance 100 Ω, output impedance 2 Ω @ ≤ 10 MHz.
(6) Slew rate 275 V/μs.

2.2.5.2 Theory of operation
AD603 is a low noise, voltage-controlled amplifier used in RF circuit andIF automatic gain control (AGC) systems. The power consumption is 125 mW at the recommended ±5 V supplies. Figure 2.10 is a simplified schematic.

The AD603 is composed of a fixed-gain amplifier, a broadband passive attenuator of 0–42.14 dB and a gain control scaling factor of 40 dB/V. The fixed gain amplifier is laser trimmed in different ranges: 31.07 dB (×35.8), 50 dB (×358) or any range in between tuned by an external resistor between Pin 5 and Pin 7. If we want get higher gain, a resistor from Pin 5 to common is needed. However, at the maximum output voltage the maximum gain is about 60 dB. For any given range, the bandwidth of the amplifier is independent of the voltage-controlled gain. The voltage-controlled amplifier provides an

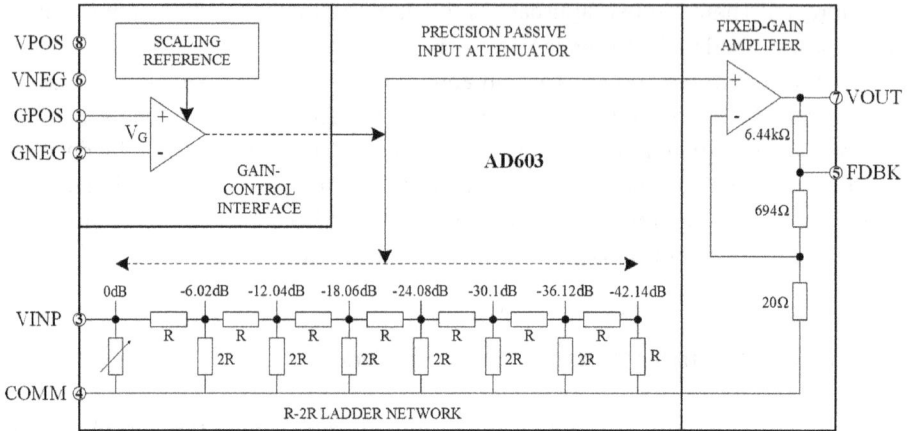

Fig. 2.10: Simplified schematic of AD603.

under-range and over-range of 1.07 dB in all cases. For example, the total gain is from −11.07 dB to +31.07 dB in the bandwidth of 90 MHz (Pin 5 and Pin 7 strapped).

The dB gain is accurately calibrated and linear, with stable over temperature and supply. The gain is controlled by high impedance (50 MΩ) and low bias (200 nA) differential input voltage. The voltage-controlled gain scale is 25 mV/dB, which means 1 V control voltage can control 40 dB gain range. It is can provide over-range and under-range of 1 dB at any selected range. For 40 dB change, the gain control response time is less than 1 μs.

AD603 can drive the 100 Ω load impedance with low distortion. For a shunt load with 500 Ω in parallel 5 pF, the total harmonic distortion is typically −60 dBc with a ± 1 V sinusoidal output at 10 MHz. The peak output is ±2.5 V minimum with a 500 Ω load.

2.2.5.3 Gain control introduction

The passive attenuation is controlled by a differential high impedance (50 MΩ) input voltage. The scaling factor is 40 dB/V or 25 mV/dB. For example, when the differential input control voltage V_G = 0 V, the attenuator offering an attenuation of 21.07 dB with slider is centered. For the maximum bandwidth 90 MHz, this leads to a total gain of − 21.07 dB + 31.07 dB = 10 dB. When the input control voltage is −500 mV, the attenuation is 21.07 dB + 0.500 V × 40 dB/V = 41.07 dB, and the overall gain is −10 dB (= −41.07 dB + 31.07 dB). When the input control voltage is +500 mV, the attenuation is 1.07 dB (= −21.07 dB + 0.500 V × 40 dB/V), and the overall gain is up to +30 dB (= −1.07 dB + 31.07 dB). When the input control of this interface exceeds limiting voltage in either direction, that is, upper voltage is over +500 mV, the gain approaches 31.07 dB (= 0 + 31.07 dB), either lower voltage less than −500 mV, the gain is as low as −11.07 dB (= −42.14 dB + 31.07 dB). In the gain control interface, the gain control voltage must be kept within the common-mode range from −1.2 V to +2.0 V with +5 V supplies. So, the basic overall gain of AD603 can be computed by

$$\text{Gain} = 40V_G + 10 \tag{2.40}$$

Where V_G is in volts, unit of gain is in dB. When Pin 5 and Pin 7 are connected together, the gain can be written as

$$\text{Gain(dB)} = \begin{cases} 40V_G + 20, \text{for 0 to} + 40\text{dB} \\ 40V_G + 30, \text{for} + 10\text{dBto} + 50\text{dB} \end{cases} \tag{2.41}$$

We can calculate the gain of the output terminal of the AD603 employing Pin 5. Figure 2.11 shows the default mode with −10 dB to +30 dB with 90 MHz bandwidth.

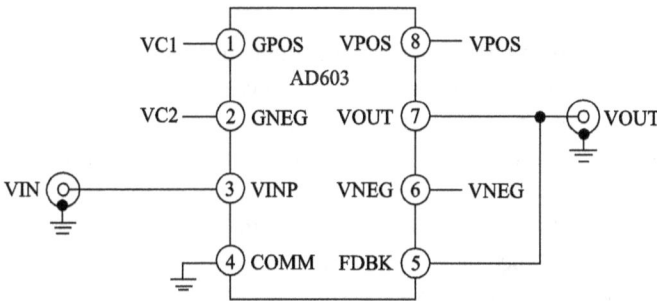

Fig. 2.11: Application circuit.

2.3 Analysis of radio frequency power amplifier

RFPA is a key module to build a radio communication system successfully. It is the biggest power consumption module in wireless transceiver. In order to reduce power consumption and extend battery life, high efficiency property is greatly desired. The output power and efficiency of the former small signal amplifier cannot meet the wireless communication requirements. Therefore, we must study other high PA technology.

RFPA is an energy converter, which transforms direct current (DC) energy from the source into high power RF energy by the control of the small power input signal. The output signal waveform and spectrum should be the same as the input signal.

A topology structure of an RFPA is shown in Fig. 2.12. It consists of a transistor such as BJT, MOSFET or MESFET, input matching network (IMN), output matching network (OMN) and RF choke. The OMN has several functions such as impedance conversion, harmonic suppression and filtering objective signal to load.

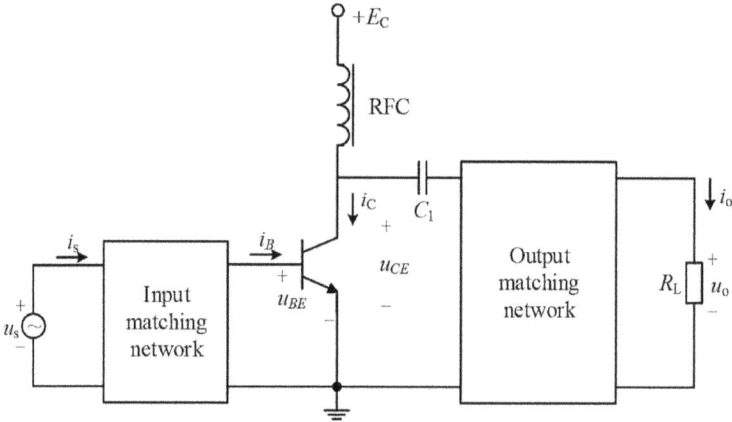

Fig. 2.12: RF power amplifier block diagram.

In an RFPA, a transistor can be operated as a dependent current source or as an electronic switch. If the PA is operated in the overdriven mode, a transistor works partially as a dependent source and partially as a switch.

From Fig. 2.15, when the BJT works as a dependent current source, the collector current waveform is determined by the base current amplitude and Q point (or quiescent point) of the transistor. Voltage waveform at the collector is determined by the dependent source and the OMN impedance. When a transistor works as a switch, its voltage is zero when the switch is ON and the current is determined by its external circuit. On the contrary, when the switch is OFF, the current is zero and its voltage is determined by its external circuit response.

2.3.1 Performance of radio power amplifier

Output power, power gain and efficiency are the basic performance indices of PAs. In a PA, the output power P_{out} is the power delivered to the external circuit of a specified frequency f. When the resonant frequency of the OMN f_0 is equal to f, P_{out} is given by

$$P_{out} = \frac{1}{2} U_m I_m = \frac{1}{2} I_m^2 R_c = \frac{1}{2} \frac{U_m^2}{R_c} \tag{2.42}$$

Where R_c is the equivalent resistance of the transistor collector, U_m is the voltage amplitude at the collector and I_m is the current amplitude at the collector.

P_L is the power delivered to the load (usually with the impedance of 50 Ω), which is

$$P_L = \frac{1}{2}\frac{U_{om}^2}{R_L}$$ (2.43)

The dissipate power of the OMN is then

$$P_N = P_{out} - P_L$$ (2.44)

The input power P_{in} is the available power to the transistor

$$P_{in} = \frac{1}{2\pi}\int_0^{2\pi} i_B u_{BE} d(\omega t)$$ (2.45)

The power gain G_p is defined by the ratio between the output and the input power

$$G_p = \frac{P_{out}}{P_{in}}$$ (2.46)

Since the broad dynamic range of the signals is handled by PAs, logarithmic unit is usually found more convenient. Assuming 1 mW as a reference, power level is compared with 1 mW and expressed in decibels, that is, in dBm

$$P_{dBm} = 10 \cdot \lg\left(\frac{P_W}{1mW}\right) = 10 \cdot \lg(P_W) + 30 \quad \text{(dBm)}$$

$$P_W = 10^{\frac{P_{dBm}}{10} - 3}\,(\text{W})$$ (2.47)

Similarly, power gain in logarithmic scale is adopted which is defined as

$$G_{dB} = 10 \cdot \lg(G_p) = P_{outdBm} - P_{indBm}$$ (2.48)

From an energy point of view, regardless of specific applications, a PA can be ultimately regarded as a device converting DC power from the supply (P_{DC}) into RF power (P_{out}). The DC power of an amplifier is

$$P_{DC} = E_c I_0$$ (2.49)

Where E_c is the source voltage and I_0 is the DC component of the output current.

Generally, the amplifier's efficiency η_c is measured by the ratio between the output RF power and the supplied DC power. Which is the effectiveness of power conversion and given as below

$$\eta_c = \frac{P_{out}}{P_{DC}} = \frac{\frac{1}{2}U_m I_m}{E_c I_0} = \frac{1}{2}\frac{U_m}{E_c}\frac{I_m}{I_0} = \frac{1}{2}\xi\gamma$$ (2.50)

Where $\xi = U_m/E_c$, which is the voltage utilization factor at the collector and $y = I_m/I_0$ is the current utilization factor at the collector. Efficiency is usually also specified as drain efficiency (η_d) or collector efficiency (η_c) in the case of a solid-state PA based on field effect or bipolar transistors, respectively.

Power dissipation at the collector is

$$P_C = \frac{1}{2\pi} \int_0^{2\pi} i_C u_{CE} d(\omega t) = P_{DC} - P_{out} \tag{2.51}$$

As the operating frequency increases, the PA power gain decreases as a result of its active constituents gain roll-off behavior. The contribution to the output power comes directly from the input drive and cannot be neglected. The input power constitutes a significant portion of the total output for signals at microwave frequencies and beyond. Therefore, the additional power P_{add} or the net increase power from the output to input terminal, is defined as

$$P_{add} = P_{out} - P_{in} \tag{2.52}$$

The power-added efficiency (PAE or η_{add}) is defined as the ratio between the added power and the supplied DC power which is a very important and highly adopted parameter of PAs.

$$PAE = \eta_{add} = \frac{P_{add}}{P_{DC}} = \frac{P_{out} - P_{in}}{P_{DC}} = \frac{P_{out} \cdot \left(1 - \frac{1}{G_p}\right)}{P_{DC}} = \eta \cdot \left(1 - \frac{1}{G_p}\right) \tag{2.53}$$

where G_p is the operating power gain.

Example 2.3: A BJT RFPA is known by P_{out} = 12 W, P_{DC} = 18 W and P_{in} = 1 W. Calculate the collector efficiency, PAE and power gain.

Solution: The collector efficiency of the PA is

$$\eta_c = \frac{P_{out}}{P_{DC}} = \frac{12}{18} = 66.7\%$$

The PAE is

$$PAE = \frac{P_{out} - P_{in}}{P_{DC}} = \frac{12-1}{18} = 61.1\%$$

The power gain is

$$G_p = \frac{P_{out}}{P_{in}} = \frac{12}{1} = 12 = 10 \lg(12) \approx 10.8(dB)$$

2.3.2 Classification of power amplifier

PAs are normally classified based on of their operating classes. When a transistor operates as a dependent-current source, the classification of an RFPA depends on the conduction angle 2θ of current at the collector. For sinusoidal input current i_b (at the base), current waveforms i_c of a BJT of various operation classes are shown in Fig. 2.13.

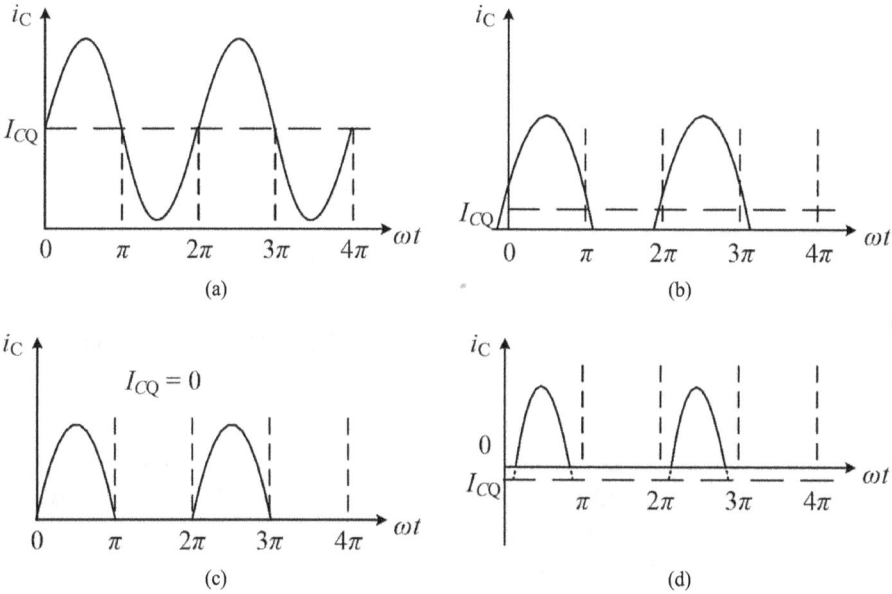

Fig. 2.13: Waveforms of current i_c in various operation classes: (a) Class A; (b) Class AB; (c) Class B; and (d) Class C.

For Class A, the conduction angle 2θ is 2π. The base-to-emitter voltage must be higher than the BJT threshold voltage U_{th} all the time. As a result, the BJT conducts during the entire cycle. The quiescent current I_{CQ} must be greater than the amplitude of the AC component I_m at the collector. The efficiency of Class A is no more than 50% because of higher quiescent collector current.

For Class AB, the conduction angle 2θ is between π and 2π. The base-to-emitter voltage is slightly higher than the BJT threshold voltage U_{th} and the BJT is biased at a small quiescent current I_{CQ}. The transistor conducts more than half a cycle but less compared to Class A. The efficiency of Class AB is higher than Class A.

For Class B, the conduction angle 2θ is π. The base-to-emitter voltage is equal to the BJT threshold voltage U_{th}. The quiescent current I_{CQ} is zero. Hence, the BJT conducts for only half a cycle. The maximum efficiency of Class B is 78.5%.

For Class C, the conduction angle 2θ is less than π. The current i_c appears for less than half a cycle. The virtual quiescent point is at the cut-off state because the biasing voltage of the base-emitter PN junction is lower than the threshold voltage U_{th} (i.e., reversely biased). The quiescent current I_{CQ} is zero. The current i_c is a series of periodic sinusoidal pulses.

The quiescent points for various class amplifiers of operation are shown in Fig. 2.14.

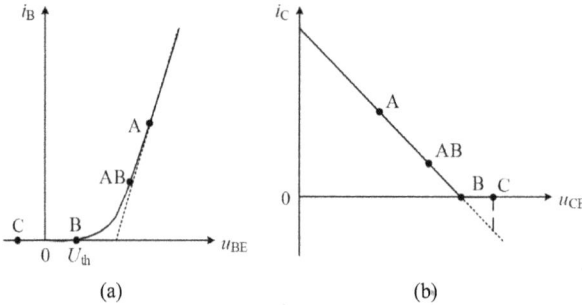

Fig. 2.14: Quiescent point for various classes: (a) input bias points and (b) output bias points.

PA such as Classes A, AB and B are usually used in low frequencies and other RF applications, whereas Class C is only used for RFPAs since it has high efficiency.

The transistor is treated as a dependent current source in Classes A, AB, B and C PAs. However, for Classes D, E and DE RFPAs, the transistor is regarded as a switch. In Class F, the transistor is partially a dependent current source and partially a switch.

2.3.3 Analysis of periodic cosine pulses

Figure 2.16(d) shows the collector current waveform of periodic cosine pulses for Class C circumstances. Here, we analyze the current waveform of the cosine pulses in detail. The waveform is shown in Fig. 2.15. The bold black line is the actual collector output current waveform. The conduction angle 2θ is less than 180°.

The subsequent analysis is applied to all cosine waveform situations. The periodic cosine pulse time-domain waveform in Fig. 2.15 can be expressed by

$$i_C = \begin{cases} I_m \cos \omega t - I_m \cos \theta , & -\theta < \omega t \le \theta \\ 0 & , & \theta < \omega t \le 2\pi - \theta \end{cases} \tag{2.54}$$

$$I_{c\,max} = I_m (1 - \cos \theta) \tag{2.55}$$

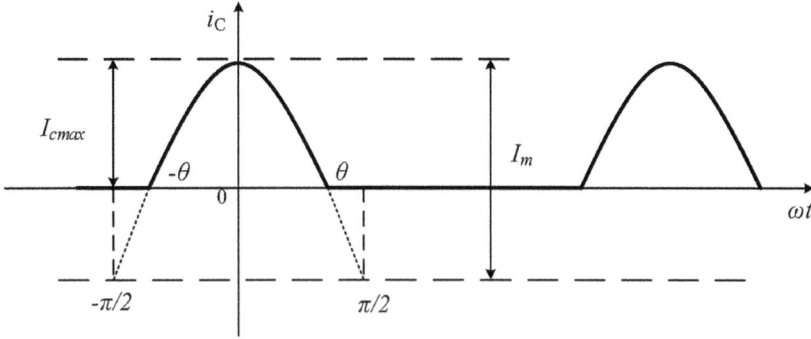

Fig. 2.15: Cosine current pulse waveform.

From eqs. (2.54) and (2.55), i_C can be rewritten as

$$i_C = \begin{cases} I_{c\max} \dfrac{\cos \omega t - \cos \theta}{1 - \cos \theta}, & -\theta \le \omega t \le \theta \\ 0, & \theta < \omega t \le 2\pi - \theta \end{cases} \tag{2.56}$$

The cosine waveform is an even function of ωt and satisfies $i_C(-\omega t) = i_C(\omega t)$. The waveform can be expanded by the Fourier series

$$i_C(\omega t) = I_{c\max}\left[\alpha_0 + \sum_{n=1}^{\infty} \alpha_n \cos n\omega t\right] \tag{2.57}$$

The DC component of the cosine-pulse current can be obtained by

$$I_{c0} = \frac{1}{2\pi}\int_{-\theta}^{\theta} i_C d(\omega t) = \frac{1}{\pi}\int_{0}^{\theta} i_C d(\omega t) = \frac{I_{c\max}}{\pi}\int_{0}^{\theta}\frac{\cos \omega t - \cos \theta}{1 - \cos \theta}d(\omega t)$$

$$= I_{c\max}\frac{\sin \theta - \theta \cos \theta}{\pi(1 - \cos \theta)} = \alpha_0 I_{c\max} \tag{2.58}$$

where

$$\alpha_0 = \frac{I_{c0}}{I_{c\max}} = \frac{\sin \theta - \theta \cos \theta}{\pi(1 - \cos \theta)} \tag{2.59}$$

The amplitude of the fundamental component of the current is

$$I_{c1m} = \frac{1}{\pi}\int_{-\theta}^{\theta} i_C \cos(\omega t)d(\omega t) = \frac{2}{\pi}\int_{0}^{\theta} i_C \cos(\omega t)d(\omega t)$$

$$= \frac{2I_{c\max}}{\pi}\int_{0}^{\theta}\frac{\cos \omega t - \cos \theta}{1 - \cos \theta}\cos(\omega t)d(\omega t) \tag{2.60}$$

$$= I_{c\max}\frac{\theta - \sin \theta \cos \theta}{\pi(1 - \cos \theta)} = \alpha_1 I_{c\max}$$

where

$$\alpha_1 = \frac{I_{c1m}}{I_{cmax}} = \frac{\theta - \sin\theta\cos\theta}{\pi(1 - \cos\theta)} \tag{2.61}$$

The amplitude of the nth harmonic is

$$
\begin{aligned}
I_{cnm} &= \frac{1}{\pi}\int_{-\theta}^{\theta} i_C\cos(n\omega t)d(\omega t) = \frac{2}{\pi}\int_{0}^{\theta} i_C\cos(n\omega t)d(\omega t) \\
&= \frac{2I_{c\,max}}{\pi}\int_{0}^{\theta}\frac{\cos\omega t - \cos\theta}{1 - \cos\theta}\cos(n\omega t)d(\omega t) \\
&= I_{c\,max}\frac{2}{\pi}\frac{\sin n\theta\cos\theta - n\sin\theta\cos n\theta}{n(n^2 - 1)(1 - \cos\theta)} = \alpha_n I_{cmax}
\end{aligned}
\tag{2.62}
$$

where,

$$\alpha_n = \frac{I_{cnm}}{I_{c\,max}} = \frac{2}{\pi}\frac{\sin n\theta\cos\theta - n\sin\theta\cos n\theta}{n(n^2 - 1)(1 - \cos\theta)} \qquad n = 2,3,4,\dots \tag{2.63}$$

The ratio between the amplitude of the fundamental component and the DC component of the current is called current utilization factor and calculated as

$$\gamma_1 = \frac{I_{c1m}}{I_{c0}} = \frac{\alpha_1 I_{cmax}}{\alpha_0 I_{cmax}} = \frac{\alpha_1}{\alpha_0} = \frac{\theta - \sin\theta\cos\theta}{\sin\theta - \theta\cos\theta} \tag{2.64}$$

From eqs. (2.56) and (2.58), we can obtain

$$i_D = I_{c0}\frac{\pi(\cos\omega t - \cos\theta)}{\sin\theta - \theta\cos\theta} \qquad -\theta \le \omega t \le \theta \tag{2.65}$$

The Fourier coefficients α_n and current utilization factor γ_1 are functions of the conduction angle θ of the current, which are shown in Fig. 2.16.

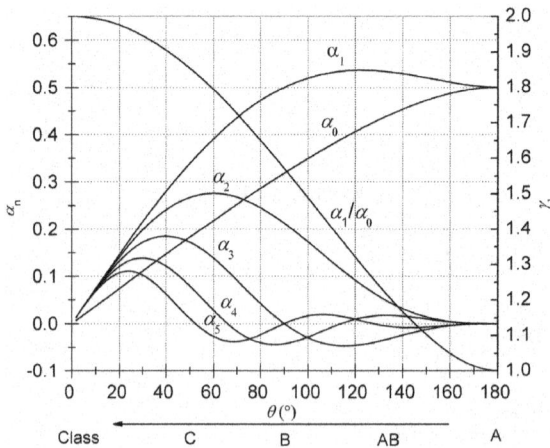

Fig. 2.16: Fourier coefficients α_n and current utilization factor γ_1.

Various coefficients for cosine-pulse waveform are given in Table 2.1. Data in the Table 2.1 is consistent with the curves in the Fig. 2.19.

Table 2.1: Coefficients for cosine-pulse waveform.

θ	α_0	α_1	γ_1	θ	α_0	α_1	γ_1
5°	0.0185	0.0370	1.9985	95°	0.3340	0.5109	1.5297
10°	0.0370	0.0738	1.9939	100°	0.3493	0.5197	1.4880
15°	0.0555	0.1102	1.9863	105°	0.3642	0.5266	1.4460
20°	0.0739	0.1461	1.9758	110°	0.3786	0.5316	1.4040
25°	0.0923	0.1811	1.9623	115°	0.3926	0.5348	1.3623
30°	0.1106	0.2152	1.9460	120°	0.4060	0.5363	1.3210
35°	0.1288	0.2482	1.9269	125°	0.4188	0.5364	1.2806
40°	0.1469	0.2799	1.9051	130°	0.4310	0.5350	1.2413
45°	0.1649	0.3102	1.8808	135°	0.4425	0.5326	1.2035
50°	0.1828	0.3388	1.8540	140°	0.4532	0.5292	1.1675
55°	0.2005	0.3658	1.8249	145°	0.4631	0.5250	1.1337
60°	0.2180	0.3910	1.7936	150°	0.4720	0.5204	1.1025
65°	0.2353	0.4143	1.7604	155°	0.4800	0.5157	1.0744
70°	0.2524	0.4356	1.7253	160°	0.4868	0.5110	1.0498
75°	0.2693	0.4548	1.6886	165°	0.4923	0.5068	1.0294
80°	0.2860	0.4720	1.6505	170°	0.4965	0.5033	1.0137
85°	0.3023	0.4870	1.6112	175°	0.4991	0.5009	1.0036
90°	0.3183	0.5000	1.5708	180°	0.5000	0.5000	1.0000

2.4 Radio frequency tuned Class C power amplifier

The conduction angle of Class C PA is less than 90°. Collector output current waveform of the transistor is a series of periodic peak cosine pulses. In order to obtain the complete cosine voltage waveform on the load, a resonant network or tuned filter should be used between the collector of the transistor and the load. This is known as the OMN of the PA, which is not only for impedance conversion purpose but also the fundamental of the input signal to the load. Though the current waveform of the transistor exists for less than half a cycle, the voltage waveform is the entire cosine waveform because of the resonant circuit. Figure 2.17 is a schematic circuit of Class C PA. E_C and E_B are biasing voltages at the collector and the base respectively. The π-shaped decoupling circuit composed of C_1, C_2 and RFC is to prevent RF current from affecting the power supply $+ E_C$, capacitor C_B provides high-frequency bypass path to ground for power supply E_B. LC resonant circuit filters the fundamental frequency for the load R_L.

Class C PA usually operates with large input signals and in nonlinear status. Graphic method is a very useful way for analysis. In order to simplify the nonlinear

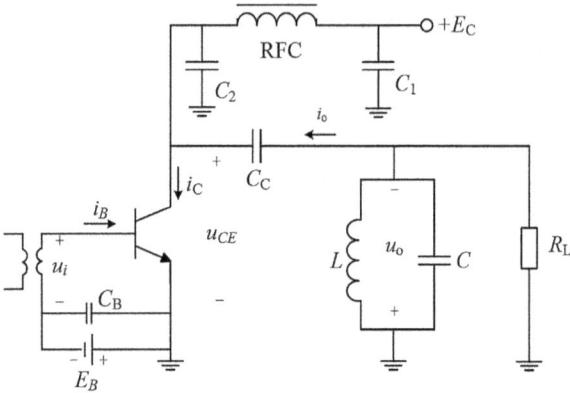

Fig. 2.17: Schematic of Class C power amplifier.

transistor characteristic while maintaining the physical meanings, the piecewise-linear approximation technique will be discussed.

2.4.1 Piecewise-linear approximation technique

The piecewise-linear approximation is a technique of replacing the curved or non-linear parameter dependence by proper segmented straight lines. Figure 2.18 shows how we can use this technique for the input and output characteristic of the transistor respectively.

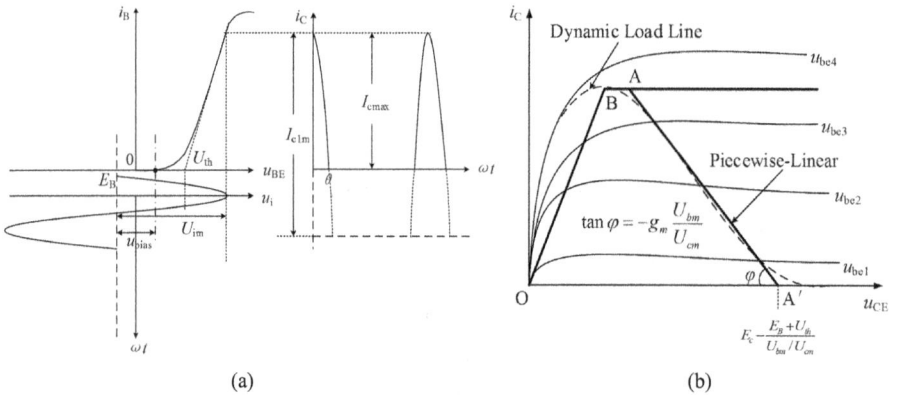

(a)

(b)

Fig. 2.18: Piecewise-linear approximation of transistor: (a) input characteristic of transistor and (b) output characteristic of transistor.

The function of how the emitter current i_B responds to u_{BE} is approximated by two dashed straight lines in Fig. 2.18(a). In large signal operating mode, the waveforms corresponding to these two dependencies are practically the same for the most part. The pinch-off region of the transistor leads to serious nonlinear distortion and is applied only for switch mode. In fact, the first two output current components, DC and fundamental, can be calculated through a Fourier series expansion with sufficient accuracy. As a result, to a linear PA, such a piecewise-linear approximation can be effective for a quick estimate of the output power and efficiency.

The dashed line in Fig. 2.18(b) is the dynamic load line (DLL) of the RF resonant PA. The DLL is the dynamic trail of the operating point codetermined by current i_c and voltage u_{ce} within a signal cycle. The DLL is determined both by the characteristic of transistor and external circuit of transistor. The actual DLL is not a straight line because of the nonlinearity of transistor. It is very difficult to analyze the actual DLL. Similar to Fig. 2.18(a), the DLL can be approximated by straight lines shown in Fig. 2.18(b). Here is $i_c - f(u_{be}, u_{ce})$, which is different from low frequency transistor output characteristic curve $i_c - f(i_b, u_{ce})$.

In order to get the mathematical expression of the cosine pulse i_c, we now start from the piecewise-linear characteristic and deduce

$$i_c = \begin{cases} g_m(u_{BE} - U_{th}) & u_{BE} \geq U_{th} \\ 0 & u_{BE} < U_{th} \end{cases} \tag{2.66}$$

Where, g_m is the transistor transconductance and U_{th} is the threshold voltage.

The input includes two parts, DC biasing voltage and AC. We assume the AC signal is in the cosine form and the input is given by

$$u_{BE} = -E_B + U_{bm} \cos \omega t \tag{2.67}$$

When the input $u_{BE}(\omega t)$ is equal to the threshold voltage U_{th} the output current $i_C(\omega t)$ is 0 (this moment defines the conduction angle $\omega t = \theta$).

$$U_{th} = -E_B + U_{bm} \cos \theta \tag{2.68}$$

where, the positive E_B is the base biasing voltage and U_{bm} is the amplitude of the input voltage. The conduction angle θ can be calculated as

$$\cos \theta = \frac{E_B + U_{th}}{U_{bm}} \tag{2.69}$$

As for the external circuit of the transistor, the output voltage is known as

$$u_{CE} = E_C - u_o(t) = E_C - U_{c1m} \cos \omega t \tag{2.70}$$

where, $u_o(t) = i_c r_c$ and r_c is the equivalent resistance of the LC resonant circuit with load. Combining eqs. (2.67), (2.68) and (2.70), we can get

$$i_C = -g_m \frac{U_{bm}}{U_{c1m}} \left[u_{CE} - \left(E_C - \frac{E_B + U_{th}}{U_{bm}/U_{c1m}} \right) \right] \quad (2.71)$$

In Fig. 2.21(b), the slope of DLL's linear region (AA' section) is $\tan \varphi = -g_m \frac{U_{bm}}{U_{c1m}}$ with the horizontal axis intercept of $E_C - \frac{E_B + U_{th}}{U_{bm}/U_{c1m}}$. The saturated region of the DLL will be discussed later.

In Fig. 2.21 (a), the collector current waveform i_C can be given as

$$i_C = I_{c1m}(\cos \omega t - \cos \theta) \quad (2.72)$$

When $\omega t = 0$, then $i_C = I_{max}$ and

$$I_{c\,max} = I_{c1m}(1 - \cos \theta) \quad (2.73)$$

The collector current i_C can be written as eq. (2.56).

2.4.2 Operating principle of Class C power amplifier

The principle of Class C PAs is illustrated in Fig. 2.19. In the linear region of the transistor, the PN junction voltage u_{be} starts increasing as the cosinoidal input signal voltage rises (QA'). Once the transistor is conducting, the dynamic operating point of the transistor will rise along with the DLL (A'B) and the output current i_c also increases.

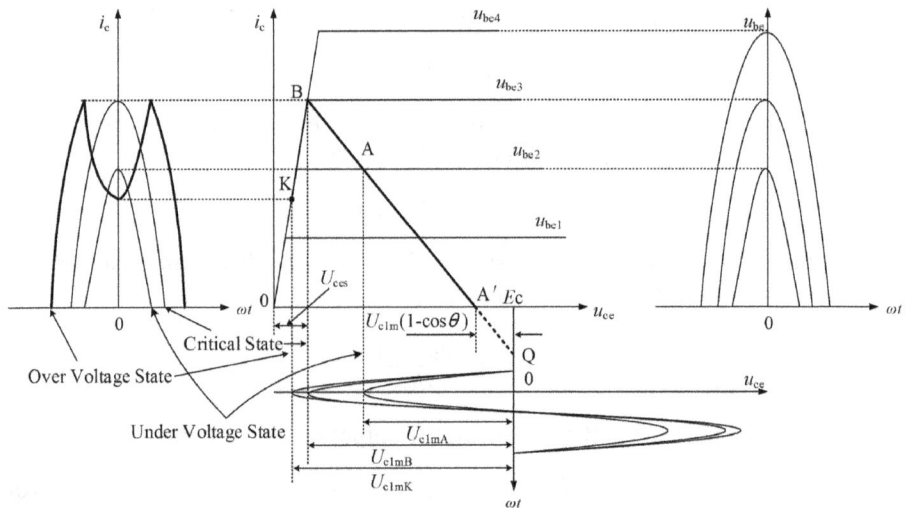

Fig. 2.19: Class C power amplifier operation process.

If the maximum of the input voltage u_{be} is lower than u_{be3} in Fig. 2.19 (such as point A at u_{be2}), the transistor stays working in the linear region and has not entered the saturation state. The collector output current waveform is pulses with full-shaped cosinoidal peaks. This state is known as the undervoltage state (UVS). The output voltage u_{ce} is comparatively low in UVS and the peak cosinoidal pulse waveform has no distortion.

When the collector current increases along line A'B of the linear DLL, u_{ce} amplitude will increase but its minimum value gets smaller. When the input voltage u_{be} makes the transistor pinched-off and thus the collector current i_c zero, the output voltage waveform peak will be flatted if the load is purely resistive. On the contrary, we know that inductor and capacitor can store and transform energy during each cycle. Hence, the output voltage waveform is in complete cosinoidal shapes because of the LC resonant circuit load. In Fig. 2.22 the maximum u_{ce} can be given

$$u_{cemax} = E_C + U_{c1m} \tag{2.74}$$

The largest amplitude of the output fundamental voltage is

$$U_{c1m} = E_C - U_{ces} \tag{2.75}$$

where, U_{ces} is the saturated voltage of the transistor which is usually a small value. If a transistor is ideal and $U_{ces} = 0$ at saturation, the maximum possible u_{ce} is $2E_C$.

If input u_{be} increases and the dynamic point moves from A to B, the transistor reaches the boundary of the linear state. Point B is called the critical state (CS). At this moment, the collector current waveform can still maintain the shape of full cosinoidal peaks and the amplitude is maximized. The output voltage u_{ce} can go down to the collector–emitter saturation voltage u_{ces}, that is, $u_{ce\ min} = u_{ces}$. The reciprocal of the slope of DLL A'AB is the dynamic or effective resistance of Class C PA

$$R'_c = \frac{U_{c1m}(1 - \cos\theta)}{I_{cmax}} = \frac{I_{c1m}R_c(1 - \cos\theta)}{I_{cmax}} = \alpha_1(\theta)R_c(1 - \cos\theta) \tag{2.76}$$

where R_c is the equivalent collector resonant load resistance given by

$$R_c = \frac{U_{c1m}}{I_{c1m}} = \frac{E_C - U_{ces}}{\alpha_1(\theta)I_{cmax}} = \frac{1}{2}\frac{U_{c1m}^2}{P_{out}} = \frac{(E_C - U_{ces})^2}{2P_{out}} \tag{2.77}$$

If the input voltage increases further, the transistor becomes saturated. During saturation, voltage u_{ce} is lower than u_{ces}. The ability of the collector to attract electrons is weakened compared with the base, and the collector current starts to drop quickly. Dynamic operating point B is heading to point K. The resulting i_C waveform has funnel-shaped depression on every peak. This state is the overvoltage state (OVS).

2.4.2.1 Amplifier characteristic

For a Class C PA, when the input excitation voltage U_{bm} is small, the transistor operates in the linear (or amplifying) region and the amplifier is in UVS. Collector output voltage amplitude, current amplitude, and output power increases as the input voltage increases. Since the conduction angle is changing, the change of the output signal parameters is nonlinear.

If the input voltage U_{bm} increases and the transistor reaches to CS, the output voltage (fundamental frequency) amplitude, current amplitude and output power will reach the maximum value.

If the input voltage U_{bm} further increases to saturation sate and the PA operates in OVS, the output voltage increases slowly and the output current waveform distortion occurs. This explains slow increase of the output power after saturation. The above analysis of amplifier characteristic of Class C PA is also shown in plots of Fig. 2.20.

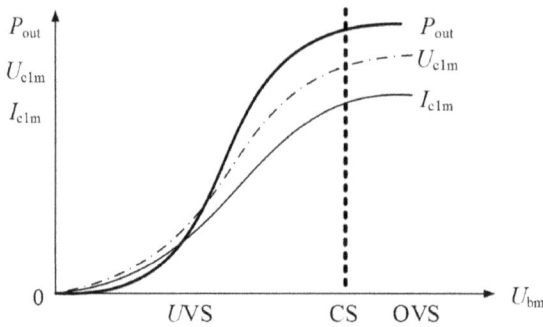

Fig. 2.20: Amplifier characteristic.

2.4.2.2 Load characteristic

By assuming that E_B, E_C and U_{bm} are fixed, we tune only one parameter, the collector equivalent load resistance R_c, and study the load characteristic of the amplifier. R_c is proportional to the reciprocal of DLL's slope. When R_c is very small, the DLL is a steep line suggesting the system is in the linear (or amplifying) region in Fig. 2.19. As R_c starts to increase, the slope of DLL (A'AB) decreases, and points A and B get closer. The output voltage amplitude U_{c1m} depends on the horizontal separation between point A and E_C, which increases rapidly as point A moves towards left. Meanwhile, I_{c1m} and I_{c0} values depend on the vertical position of point A. Since point A moves along a slightly declining line determined by the same $u_{be\ max}$, the pulse peak I_{cmax} only drops slightly as R_c increases. Therefore, the proportional I_{c1m} and I_{c0} decrease only at a very slow rate and can be considered as nonvarying in the following power and efficiency analysis. Consequently, the output power $P_{out} = \frac{1}{2} U_{c1m} I_{c1m}$ increases rapidly because of U_{c1m}; the DC source power $P_{DC} = E_C I_{c0}$ decreases very slowly because E_C is constant; the power of transistor dissipation $P_C = P_{DC} - P_{out}$ decreases quickly and the efficiency $\eta_c = P_{out}/P_{DC}$ increases fast. The changes of parameters are plotted in Fig. 2.21. The output power P_{out} reaches maximum at the CS. As R_c increases further and the amplifier is in OVS, all mentioned parameters decrease except for the output voltage U_{c1m}.

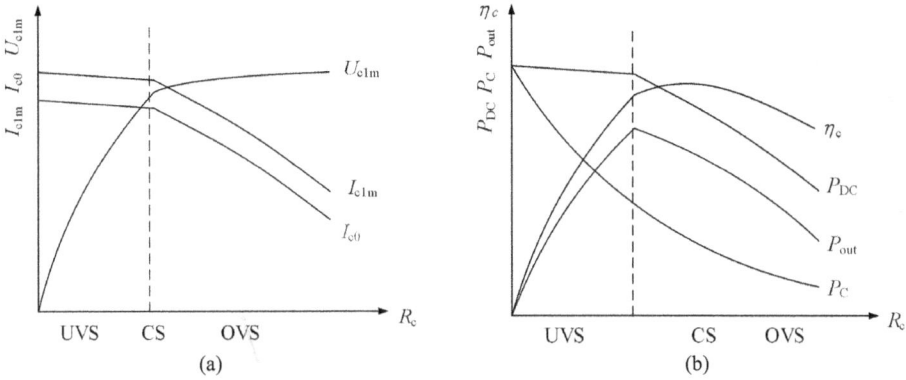

Fig. 2.21: Load characteristic, (a) voltage and current characteristic, (b) power characteristic.

2.4.2.3 Collector modulation characteristic

When the source power E_C increases the DLL A'AB will shift towards right and everything else stays unchanged in Fig. 2.19. The operating state tends to change from OVS to UVS as E_C increases. This is the foundation of collector modulation characteristic of Class C PA. The output voltage and current changes are shown in Fig. 2.22. As the system starts to leave the OVS region, the suppressed I_{cmax} quickly restores to its full height, and I_{c1m} and I_{c0} increase rapidly. U_{c1m} being the product of I_{c1m} and R_c also increases rapidly as E_C increases. This characteristic can be applied to the amplitude modulation (AM) system by combining E_C with

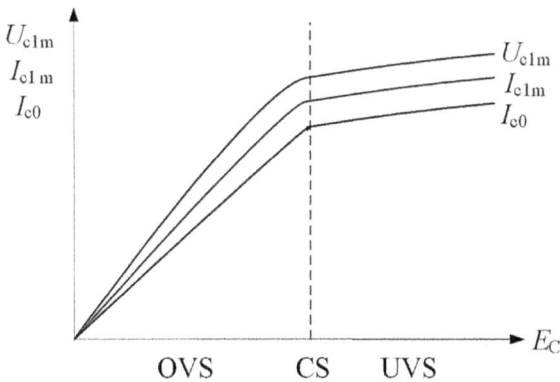

Fig. 2.22: Collector modulation characteristic of transistor.

the modulation voltage to generate AM signal. Once after the critical point and in the UVS, this modulation effect is not optimal since current and voltage are not sensitive to E_C variations.

2.4.2.4 Base modulation characteristic of Class C power amplifier

Base modulation characteristic is similar to that of amplifier characteristic because source power E_B is in series with the modulation signal U_{bm}. The mechanism is already shown in Fig. 2.19 and the external circuit characteristic are shown in Fig. 2.23. In UVS, E_B is in series with the modulation voltage, U_{c1m} increases as E_B increases. Base AM is better realized in UVS than in OVS.

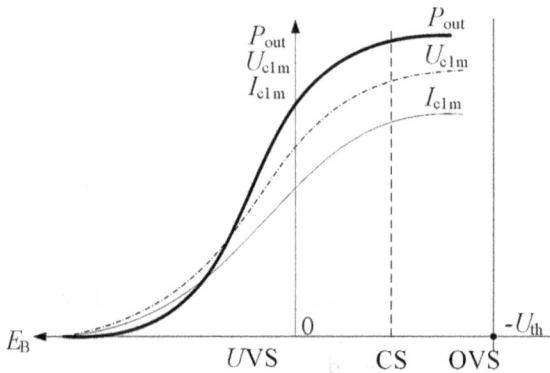

Fig. 2.23: Base modulation characteristic.

Example 2.4: A Class C PA is designed to deliver power of 2 W. The conduction angle is $\theta = 70°$ and the power supply voltage is $E_C = 12$ V. The saturation voltage of the transistor is $U_{ces} = 1$ V. Please find out the equivalent collector load resistance R_c, the consumed dc power P_{DC} and the collector efficiency η_c.

Solution: The maximum possible amplitude of the output voltage is

$$U_{c1m} = E_C - U_{ces} = 12 - 1 = 11\,(\text{V})$$

From eq. (2.77), the equivalent collector load resistance is then

$$R_c = \frac{U_{c1m}^2}{2P_{out}} = \frac{11^2}{2 \times 1.5} = 40.3\,(\Omega)$$

The amplitude of the fundamental current is

$$P_{out} = \frac{1}{2}U_{c1m}I_{c1m} \Rightarrow I_{c1m} = \frac{2P_{out}}{U_{c1m}} = \frac{2 \times 1.5}{11} = 273\,(\text{mA})$$

From eqs. (2.59) and (2.61), $\alpha_0(70°)$ and $\alpha_1(70°)$ are

$$\alpha_0(70°) = \frac{I_{c0}}{I_{c\,max}} = \frac{\sin\theta - \theta\cos\theta}{\pi(1 - \cos\theta)} = 0.2524$$

$$\alpha_1(70°) = \frac{I_{c1m}}{I_{cmax}} = \frac{\theta - \sin\theta\cos\theta}{\pi(1 - \cos\theta)} = 0.4356$$

The maximum of the collector current is

$$I_{cmax} = \frac{I_{c1m}}{\alpha_1(70°)} = \frac{273}{0.4356} = 627\,(\text{mA})$$

DC component of the collector current is

$$I_{c0} = \alpha_0(70°)I_{cmax} = 0.2524 \times 626.7 = 158\,(\text{mA})$$

DC power of the source supply is

$$P_{DC} = I_{c0}E_C = 0.158 \times 12 = 1.9\,(\text{W})$$

The collector efficiency is

$$\eta_c = \frac{P_{out}}{P_{DC}} = \frac{1.5}{1.9} = 78.9\%$$

Example 2.5: Design a Class C PA to deliver power of 5 W at $f = 100$ MHz with bandwidth of 5 MHz. The conduction angle is $\theta = 60°$ and the power supply voltage is $E_C = 12$ V. The saturation voltage of the transistor is $U_{ces} = 1$ V.

Solution: The maximum amplitude of the output voltage is

$$U_{c1m} = E_C - U_{ces} = 12 - 1 = 11\,(\text{V})$$

The equivalent collector load resistance is

$$R_c = \frac{U_{c1m}^2}{2P_{out}} = \frac{11^2}{2 \times 5} = 12.1\,(\Omega)$$

The amplitude of the output current is

$$I_{c1m} = \frac{U_{c1m}}{R_c} = \frac{11}{12.1} = 0.91\,(\text{A})$$

From eq. (2.64) and Table 2.1, the DC current is

$$I_{c0} = \frac{I_{c1m}}{\gamma_1} = \frac{0.91}{1.7936} = 0.51\,(\text{A})$$

For $\theta = 60°$, the maximum collector current is

$$I_{cmax} = \frac{I_{c0}}{\alpha_0(60°)} = \frac{0.51}{0.218} = 2.33\,(\text{A})$$

The maximum collector-to-emitter voltage is

$$U_{cem} = 2E_C = 2 \times 12 = 24\,(\text{V})$$

DC power of the supply is

$$P_{DC} = E_c I_{c0} = 12 \times 0.51 = 6.12 \,(\text{W})$$

Power dissipation at the collector is

$$P_C = P_{DC} - P_{out} = 6.12 - 5 = 1.12 \,(\text{W})$$

The collector efficiency is

$$\eta_c = \frac{P_{out}}{P_{DC}} = \frac{5}{6.12} = 81.7\%$$

The quality factor with load is

$$Q_L = \frac{f}{BW} = \frac{100}{5} = 20$$

The reactance of the resonant circuit component L is

$$X_L = X_C = \frac{R}{Q_L} = \frac{12.1}{20} = 0.61 \,(\Omega)$$

Yielding

$$L = \frac{X_L}{\omega} = \frac{0.61}{2\pi \times 100 \times 10^6} = 0.97 \,(\text{nH})$$

and

$$C = \frac{1}{\omega X_C} = \frac{1}{2\pi \times 100 \times 10^6 \times 0.61} = 2.61 \,(\text{nF})$$

2.4.3 Bias circuits

2.4.3.1 Collector bias circuit
The function of the collector bias circuit is to supply DC power to the transistor. There are two types of bias circuits, series fed and parallel fed, which are shown in Fig. 2.24.

Figure 2.24(a) shows a series-fed circuit with the transistor, the load and the supply power connected in series. Figure 2.24(b) shows a parallel-fed circuit, with those three parts in parallel. One end of the load and matching network is grounded.

2.4.3.2 Base bias circuit
The function of base bias is to provide DC voltage to the base of the transistor. Figure 2.25(a) is a zero bias circuit. The base is DC grounded by the inductor. The bias voltage is dependent on the threshold voltage of the transistor. Figure 2.25(b) is a base bias circuit. Working together with a capacitor, the voltage of the base resistance R provides the bias voltage for the transistor. In Fig. 2.25(c), the voltage of the emitter resistance R provides the bias for the transistor. By choosing appropriate R and C values, u_{BE} can be maintained automatically replacing an independent power source E_b.

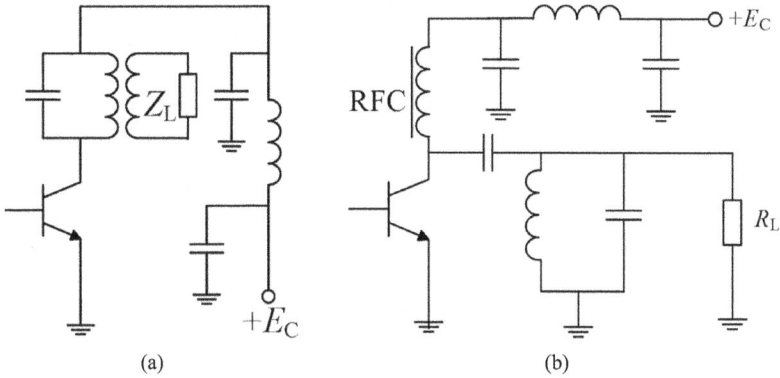

Fig. 2.24: Collector biasing circuits: (a) series-fed circuit and (b) parallel-fed circuit.

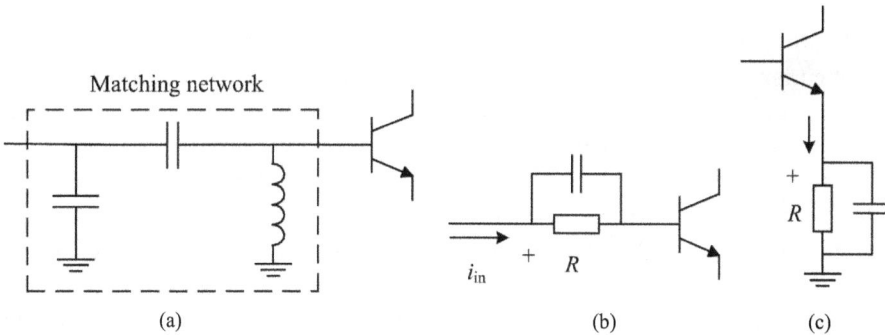

Fig. 2.25: Base bias network: (a) zero bias; (b) base bias; and (c) emitter bias.

2.5 Class E power amplifier

Class E PA is based on the hypothesis that the transistor is operated as a switch, differently from the usual current source mode for Class A, AB, B and Class C PA. The basic circuit topology of a Class E PA is shown in Fig. 2.26(a), where C_o is output capacitance of the active device. L_2 and C_2 constitute a series resonant circuit and have high Q value. Collector choke inductor L_1 has to be high enough to ensure flowing current I_{CC} keep constant. OMN is output impedance matching network to deliver the maximum power from transistor to load. In ideal state, the collector voltage is zero when the active device is in its ON state. On the contrary, the collector voltage is Ec when the transistor is in its OFF state. Collector of current and voltage waveforms have no overlap on the transistor. So, the efficiency is 100% ideally. Collector peak voltage of a Class E PA is approximately 3.56 E_c. It is worthwhile to remark that there should be higher breakdown voltage when selecting active device.

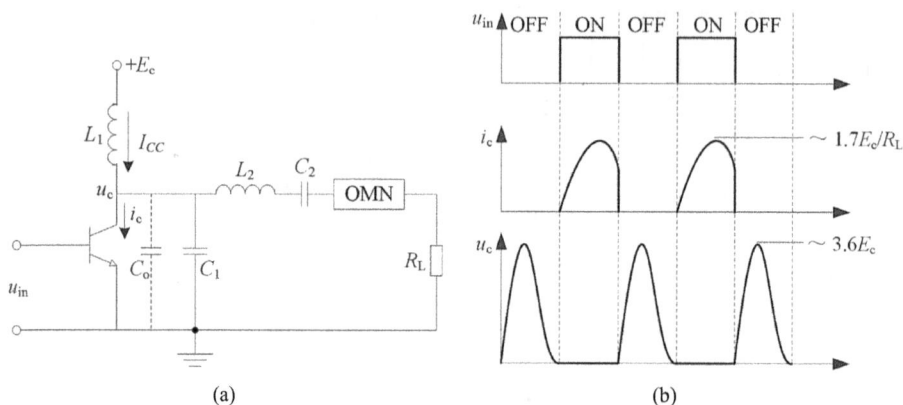

Fig. 2.26: (a) Circuit schematic of a Class E power amplifier and (b) ON–OFF states of the transistor.

Unfortunately, the active device is not an ideal switch actually. There are overlaps of current and voltage waveforms. Then the transistor will dissipate power. The collector current will not immediately be zero when the device turns off and there is a certain amount of rise time when the device turns on. To alleviate the effect of the device turns off state, a capacitor C_1 is shunted across the device to slow down the collector voltage rise. To solve the problem of turning on the device, L_2 and C_2 resonant frequencies are settled slightly less than the operating frequency. The L_2 and C_2 series tuned circuit will be equivalent to an inductor, then the voltage ahead of current is in a certain phase. This will make the collector voltage drop to zero before the device turns to the on state. The waveforms of voltage and current are depicted in Fig. 2.26(b). To switch mode device, the following conditions have to be ensured:
(1) the voltage across the device has to be zero while current is flowing (i.e., zero voltage switching, ZVS condition);
(2) the current across the device has to be zero whenever a nonzero voltage drop is present;
(3) the slope of the voltage across the device has to be zero when the device switches on.

C_1, C_2 and L_2 of Class E PA can be given by:

$$u_{ce\,max} = 3.562E_c \tag{2.78}$$

$$R_L = 0.577\frac{(E_c - U_{ces})^2}{P_o} \tag{2.79}$$

$$I_{CC} = \frac{E_c - U_{ces}}{1.734R_L} \tag{2.80}$$

$$C_1 = \frac{2}{\pi\,(1+\pi^2/4)\omega R_L}\,\frac{1}{} \simeq \frac{0.1836}{\omega R_L} \tag{2.81}$$

$$C_2 \simeq C_1 \frac{5.447}{Q}\left(1 + \frac{1.42}{Q-2.08}\right) \tag{2.82}$$

$$L_2 = \frac{QR_L}{\omega} \tag{2.83}$$

where u_{cemax} is the maximum voltage across the transistor, P_o is output power, Q value determined by the bandwidth. One uses OMN to transfer the load to R_L if R_L is not equal to expected load.

Example 2.6: Design a Class E PA operating at 900 MHz with P_o = 2 W. Known are E_c = 3.6 V, U_{ces} = 0.2 V and Q = 7.

Solution: The maximum voltage across the transistor is

$$u_{ce\,max} = 3.562E_c = 3.562 \times 3.6 = 12.8\,\text{V}$$

The load resistance is

$$R_L = 0.577 \times \frac{(3.6-0.2)^2}{1.5} = 4.45\,\Omega$$

Current of flowing the choke inductor L_1 is

$$I_{CC} = \frac{E_c - U_{ces}}{1.734R_L} = \frac{3.6-0.2}{1.734 \times 4.45} = 0.441\,\text{A}$$

The inductance of the series resonant circuit is

$$L_2 = \frac{QR_L}{\omega} = \frac{7 \times 4.45}{2\pi \times 900 \times 10^6} = 5.51\,\text{nH}$$

The shunt capacitance is

$$C_1 = \frac{0.1836}{\omega R_L} = \frac{0.1836}{2\pi \times 900 \times 10^6 \times 4.45} = 7.30\,\text{pF}$$

The capacitance of the series resonant circuit is

$$C_2 \approx C_1 \frac{5.447}{Q}\left(1 + \frac{1.42}{Q-2.08}\right) = 7.3 \times \frac{5.447}{7} \times \left(1 + \frac{1.42}{7-2.08}\right)$$

$$= 7.30\,\text{pF}$$

2.6 Class F power amplifier

Class F PAs have proper waveform shaping network and tuned load. The collector voltage waveform of the transistor is a square wave and the current waveform is a half sinusoidal with no overlap on each other. The waveforms are shown in Fig. 2.27. So, there is no power dissipation of the transistor and the theoretical efficiency is 100%.

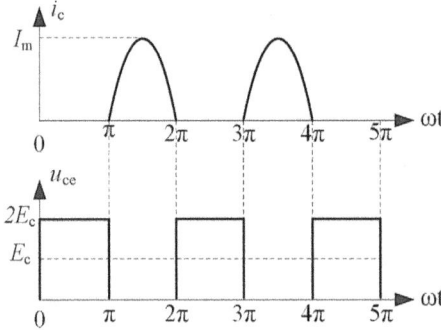

Fig. 2.27: Voltage and current waveforms of Class F power amplifier.

The load network impedances of Class F have to be given

$$Z_1 = \frac{8E_c}{\pi I_m} \tag{2.84}$$

$$Z_{2n} = 0 \quad n = 1, 2, 3, \dots \tag{2.85}$$

$$Z_{2n+1} = \infty \quad n = 1, 2, 3, \dots \tag{2.86}$$

where Z_1 is fundamental impedance, Z_{2n} is even harmonic component impedance and Z_{2n+1} is odd harmonic component impedance. Obviously, the more output harmonics, the closer to the ideal Class F waveforms in Fig. 2.27, and the higher efficiency of Class F PA. Different voltage and current harmonic combinations have different efficiencies shown in Table 2.2.

In Table 2.2, VHC is voltage harmonic components, EFF is efficiency and CHC is current harmonic components, respectively. From Table 2.2, the ideal Class F PA has maximum 75% with the second harmonic is short and the third harmonic is high impedance.

A Class F PA circuit with the third harmonic is shown in Fig. 2.28. The harmonic control network is composed of lumped components.

Table 2.2: Efficiencies of different voltage and current harmonic combinations.

VHC \ EFF \ CHC	1	1,3	1,3,5	1,3,5,7	1,3,5,. . .,∞
1	$1/2 = 0.500$	$9/16 = 0.563$	$75/128 = 0.586$	$1225/2048 = 0.598$	$2/\pi = 0.637$
1,2	$2/3 = 0.667$	$3/4 = 0.750$	$25/32 = 0.781$	$1225/1536 = 0.798$	$8/3\pi = 0.849$
1,2,4	$32/45 = 0.711$	$4/5 = 0.800$	$5/6 = 0.833$	$245/288 = 0.851$	$128/45\pi = 0.905$
1,2,4,6	$128/275 = 0.731$	$144/175 = 0.823$	$6/7 = 0.857$	$7/8 = 0.875$	$512/175\pi = 0.931$
1,2,4,. . .,∞	$\pi/4 = 0.785$	$9\pi/32 = 0.884$	$75\pi/256 = 0.920$	$1225\pi/4096 = 0.940$	1

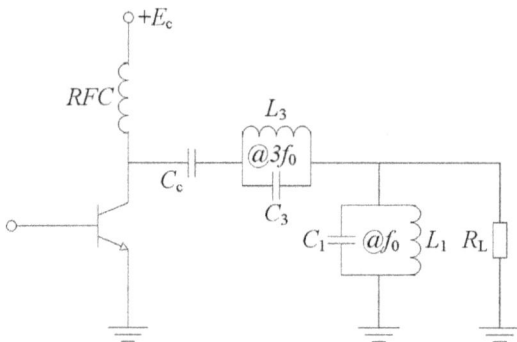

Fig. 2.28: Class F power circuit structure with the third harmonic.

Assuming that the Q value of the circuit is determined by L_1, C_1 and R_L, then

$$Q = w_0 C_1 R_L = \frac{w_0}{B_{0.7}} \tag{2.87}$$

where $B_{0.7}$ is bandwidth of the PA, C_1 becomes

$$C_1 = \frac{1}{B_{0.7} R_L} \tag{2.88}$$

Parallel tuned circuit L_1, C_1 resonant frequency is w_0, L_1 can be computed as

$$L_1 = \frac{1}{w_0^2 C_1} \tag{2.89}$$

At frequency $2f_0$, the reactance of parallel tuned circuit L_1, C_1 is negative, the reactance of parallel tuned circuit L_3, C_3 is positive, respectively. So, values of the capacitance C_1 and C_3 can be set to shunt from transistor collector to ground, that is

$$-\frac{1}{2w_0 C_1} + \frac{2w_0 L_3}{1 - (2w_0)^2 L_3 C_3} + \frac{2w_0 L_1}{1 - (2w_0)^2 L_1 C_1} = 0 \tag{2.90}$$

where

$$L_3 C_3 = \frac{1}{9 w_0^2} \tag{2.91}$$

multiplying w_0 at both sides of eq. (2.90), using eqs. (2.89) and (2.91), eq. (2.90) can be simplified as

$$-\frac{1}{C_1} + \frac{2/(9C_3)}{1 - 4/9} + \frac{2/C_1}{1 - 4} = 0 \tag{2.92}$$

or

$$\frac{1}{C_1} = \frac{4}{5C_3} - \frac{4}{3C_3} \tag{2.93}$$

Furthermore, one can adjust C_1 and L_3, C_3 parallel tuned circuit and make the reactance from transistor to load R_L equal to zero. This can ensure that the load gets nondistortion fundamental power. Then, it follows that

$$-\frac{1}{\omega_0 C_1} + \frac{\omega_0 L_3}{1 - \omega_0^2 L_3 C_3} = 0 \tag{2.94}$$

which can be simplified as

$$\omega_0^2 L_3 (C_1 + C_3) = 1 \tag{2.95}$$

from eq. (2.93), C_1 is computed as

$$C_1 = 8C_3 \tag{2.96}$$

Substituting eq. (2.96) into eq. (2.93), it is inferred that

$$C_3 = \frac{81}{160} C_1 \tag{2.97}$$

L_3, C_3 is tuned at the third harmonic, then L_3 can be computed as

$$L_3 = \frac{1}{9\omega_0^2 C_3} \tag{2.98}$$

2.7 Summary

In this chapter, RF amplifiers including small signal tuned amplifiers and PAs are discussed. Small signal tuned amplifiers are usually used as receiver front end. In order to amplify the faint signal from antenna, the tuned amplifiers must have high gain with low noise figure. PAs are used in transmitters to amplify RF modulated signals to antenna. Transistors in PAs work in nonlinear field with Class C status. PAs must have high output power with high efficiency to increase transmission distance.

Problems

2.1 Please explain the physical meaning of f_β, f_T and f_{max}. Why is f_{max} the highest, f_T moderate and f_β the lowest? Does f_{max} vary depending on different amplifier configuration? Please analyze it.

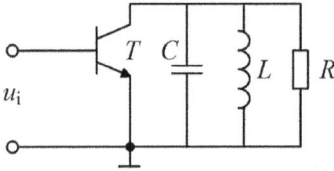

Ex.Fig. 2.1

2.2 A transistor with $f_T = 250$ MHz and $\beta_0 = 50$. Please calculate the β value at $f = 1$ MHz, 20 MHz, 50 MHz, respectively.

2.3 AC equivalent circuit of a single small signal resonant amplifier is shown in Ex.Fig. 2.1 The resonant frequency $f_0 = 10$ MHz, passband $B_{0.7} = 500$ kHz and resonant voltage gain $A_{u0} = 100$. The tested Y parameters at operating point and operating frequency are as follows:

$$y_{ie} = (2 + j0.5) \text{ mS}, \quad y_{re} \approx 0, \quad y_{fe} = (20 - j5) \text{ mS}, \quad y_{oe} = (20 + j40) \text{ μS}$$

If the quality factor of L without load $Q_0 = 60$. Please calculate resonant tank parameters L, C and external resistor R.

2.4 A single resonant amplifier with $|A_{u0}| = 20$, $B = 6$ kHz at resonant frequency. By increasing the level of the same amplifier, please calculate the sum of resonant voltage gain and that of passband. If the sum of the passband holding is 6 kHz, please calculate the single level voltage gain A_{u1}.

2.5 A resonant PA operates at critical point state if the equivalent load resistor R_c changes suddenly. If R_c is increased by double or reduced by half, find how to change of output power P_o. Please analyze the reasons of changing.

2.6 In a resonant PA, assume that $E_C = 24$ V, $I_{co} = 250$ mA, output power $P_o = 5$ W and collector voltage utilization factor $\xi = 1$. Please calculate transistor dissipate power P_c, collector equivalent load resistor R_c, collector efficiency η_c and collector current fundamental component amplitude I_{c1m}.

2.7 A resonant PA with operates at critical point state. Assume that $E_C = 24$ V, current conduction angle $\theta = 70°$, $\alpha_0 (70°) = 0.253$, $\alpha_1 (70°) = 0.436$, collector current pulse maximum value $I_{cmax} = 2.2$ A, collector voltage utilization factor $\xi = 0.9$. Please calculate output power P_{out}, power of power supply P_{DC}, collector efficiency η_c and collector equivalent load resistor R_c.

2.8 Design a Class C PA delivering power of 1 W at $f = 500$ MHz with bandwidth 25 MHz. The conduction angle is $\theta = 45°$ and the power supply voltage is $E_C = 5$ V. The saturation voltage of transistor is $U_{ces} = 0.3$ V.

Chapter 3
Sinusoidal oscillators

3.1 Introduction

An electronic circuit that produces periodic, oscillating electronic signals is called an oscillator, typically in sinusoidal, square, triangular or saw tooth waveforms. From a power supply, oscillators can convert direct current (DC) into alternating current (AC) signal without needing any AC input.

In many electronic devices, oscillators are widely used. Most commonly used oscillators include radio and television transmitters or receivers, function generators, test equipments and clock signals in computers and quartz clocks.

There are many types of oscillators. According to the operational principle, oscillators can be divided into feedback type or negative resistance type; according to waveform, oscillators can be sinusoidal or nonsinusoidal; according to the used frequency-selective filter, oscillators can be LC oscillators, crystal oscillators or RC oscillators.

Feedback-type oscillators are the most commonly used sinusoidal oscillators, while negative resistance oscillators are mainly used in the microwave band. As for feedback oscillators, LC oscillators and crystal oscillators are usually used for high-frequency sinusoidal signals, and RC oscillators are used for low-frequency sinusoidal signals.

With the emergence of modern radio and radar systems, it is necessary to provide stable harmonic oscillations at specific carrier frequencies in order to establish the required modulation and mixing conditions. Although the early carrier frequencies are mostly low- to medium-frequency band, today's RF systems can easily exceed 1 GHz point. This leads to the need for specialized oscillator circuits that can provide stable and pure sinusoidal signals.

In this chapter, we focus on feedback oscillators. Once the fundamental idea of how to generate oscillations is mastered, we investigate high-frequency feedback sinusoidal oscillators such as LC oscillators and quartz crystal oscillators. Typical LC oscillator circuits include Colpitts, Hartley, Clapp and Seiler circuits. Crystal oscillators often look similar to LC oscillators with the crystal replacing part of the resonant circuit. Pierce oscillator is a commonly used example of crystal oscillators.

The most import performance index of a sinusoidal oscillator is frequency stability. With the development of modern electronic technology, the standard of frequency stability is higher and higher. Therefore, applications with high-frequency stability are tremendous.

https://doi.org/10.1515/9783110593822-003

3.2 Operational principle of feedback oscillators

In this section, we first introduce basic principle of feedback oscillators which consist of positive-feedback amplifier to form oscillation and frequency-selective filter to decide the oscillation frequency. Then we analyze the startup, balance and stability conditions of oscillation.

3.2.1 Basic principle and analysis of oscillation

In fact, a feedback oscillator is evolved from a feedback amplifier. In Fig. 3.1, when switch K is set at point 1, the circuit is a tuned amplifier. Adjusting the mutual inductance M, homonymous ends and loop parameters, the feedback signal u_f can be finally equal to the input signal u_i. At this time, K is quickly switched to point 2. While the circuit still maintains its state, the input end is changed to u_f instantaneously. Now a tuned amplifier becomes a self-excited oscillator.

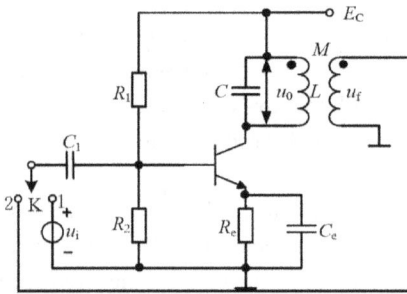

Fig. 3.1: From the tuned amplifier to the feedback oscillator.

The core of any oscillator circuit is a loop that produces positive feedback at a selected frequency. Figure 3.2 illustrates the generic closed-loop system diagram. Amplifier and feedback network are two main parts of the feedback oscillator, where A is the gain of amplifier and F is the feedback coefficient of feedback loop.

If the feedback signal u_f is equal to the input signal u_i when switch is on "1," the amplifier can continue working after being switched to "2" relying on the feedback u_f even if there is no input. The amplifier becomes an oscillator now.

Since

$$u_f = \dot{F} \cdot u_o = \dot{F} \cdot \dot{A} \cdot u_i \qquad (3.1)$$

Then

$$\dot{A} \cdot \dot{F} = 1 \qquad\qquad (3.2)$$

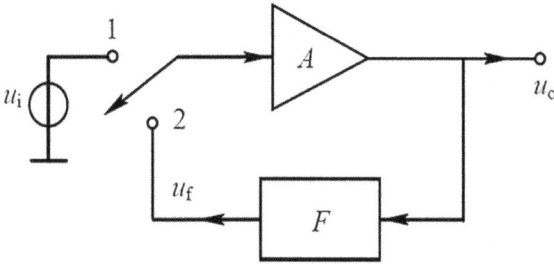

Fig. 3.2: Basic feedback oscillator configuration.

3.2.2 Oscillation conditions of feedback oscillators

3.2.2.1 Startup condition of oscillation

The most common form of feedback oscillators is an amplifier, such as a transistor or an operational amplifier, with a feedback loop going back into the input end through a frequency-selective filter to provide positive feedback. When the power supply is switched on, electronic perturbations in the circuit provide nonzero input signal to get oscillation started from the small perturbations. The small signal passes through the amplifier and the positive feedback loop for many cycles and finally builds up an oscillation signal at a fixed frequency with decent and stable amplitude.

It should be pointed out that the product of A and F should be greater than 1 at the buildup stage instead of equal to 1. Then the feedback voltage as the new input will be larger than the previous cycle every time, helping the oscillation buildup effectively. So the startup condition of oscillation is

$$\dot{A} \cdot \dot{F} > 1 \qquad\qquad (3.3)$$

The earlier condition of oscillation contains two aspects

$$\begin{cases} AF > 1 \\ \varphi_A + \varphi_F = 2n\pi \quad (n = 0, \ \pm 1, \ ...) \end{cases} \qquad (3.4)$$

The first condition is on amplitude and the second is on phase. Positive feedback is possible when they are both met.

Will the amplitude of the oscillation keep increasing forever? The answer is no. In Fig. 3.3, because of the nonlinear characteristics of amplifier, the oscillation signal u_{ce} will be automatically stabilized to a certain extent and the transistor amplification

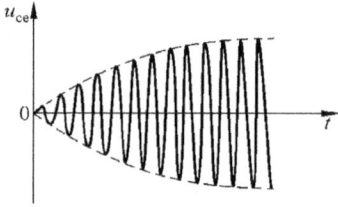

Fig. 3.3: The buildup of oscillation signal.

A is no longer as large as what it was in the buildup stage. The circuit is now in equilibrium, or what we call the balance state. This leads to the next part.

3.2.2.2 Balance condition of oscillation

When the oscillation voltage u_o stabilizes to a certain extent, the balance state is established and the feedback voltage holds constant for every cycle. So eq. (32) is the balance condition, which can be unfolded as two aspects:

$$\begin{cases} AF = 1 \\ \varphi_A + \varphi_F = 2n\pi \quad (n = 0, \ \pm 1, \ \ldots) \end{cases} \tag{3.5}$$

Here, the first condition is on amplitude and the second one is on phase. In the balance state, the energy from the power supply compensates the circuit loss while maintaining the amplitude of the output voltage unchanged.

Because of the nonlinear characteristics of amplifier, as long as the amplitude startup condition of $AF > 1$ is met, the amplitude balance condition $AF = 1$ is easy to achieve. The phase balance condition must be satisfied all the time to maintain a positive feedback.

3.2.2.3 Stability condition of oscillation

The established balance state may be destroyed due to many unstable factors in the circuit such as power supply fluctuation, temperature variation and mechanical vibration. In Fig. 3.4 we list two types of balance state. Once perturbed away from the balance position, the ball can automatically get back to its original state and then the balance state is considered stable. Otherwise the balance is unstable.

(a)   (b)

Fig. 3.4: Two types of balance state: (a) unstable balance and (b) stable balance.

The stability condition of oscillation also can be analyzed from two aspects: amplitude and phase.

(1) Amplitude stability condition: The oscillator should be capable of resisting amplitude fluctuations near equilibrium to keep amplitude stable. To be more specific, in the vicinity of equilibrium, when amplitude changes suddenly, the amplifier gain A should self-adjust so that the amplitude can vary in the opposite way and compensate the sudden increase or decrease. Thereby, the amplitude returns to the balance level instead of deviating away.

In Fig. 3.5 the graphs show different relationships between the amplifier gain and the input voltage. In Fig. 3.5(a), the amplifier gain A decreases monotonously as the input voltage rises. Point C is the balance point satisfying $AF = 1$. With the negative slope, if the input increases the gain reduces and if the input decreases the gain enlarges. Therefore, the final output amplitude can be adjusted and kept stable. However, in Fig. 3.5(b), another balance point B with positive slope shows unstable property through the similar analysis. Readers can repeat the earlier thinking procedure and naturally get to the conclusion that the amplitude stability condition is

$$\frac{dA}{du}\bigg|_{A=\frac{1}{F}} < 0 \tag{3.6}$$

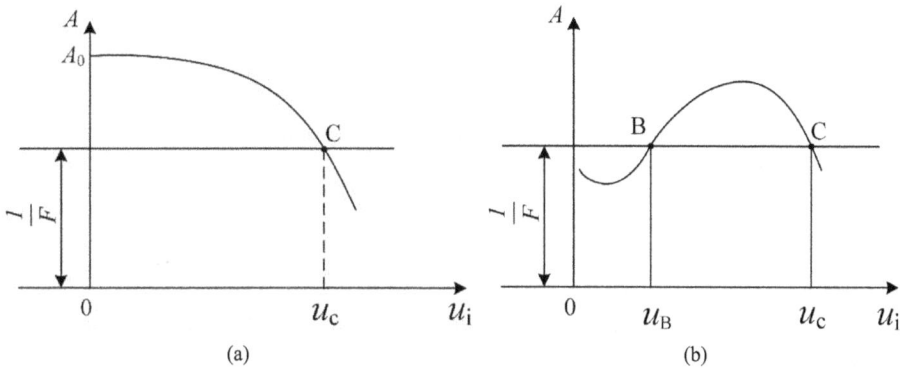

Fig. 3.5: Relationship between the amplifier gain and the input voltage. (a) balance point, (b) balance and unbalance points.

(2) Phase stability condition: Phase stability and frequency stability is essentially the same thing, because phase change inevitably brings about frequency change. Phase stability condition requires that when the phase is away from the balance point, the circuit itself can restore the original frequency or phase. Similarly, in Fig. 3.6 we use the phase–frequency characteristic curve to illustrate. The slope of the phase–frequency characteristic curve in the vicinity of the operating frequency is negative in this plot.

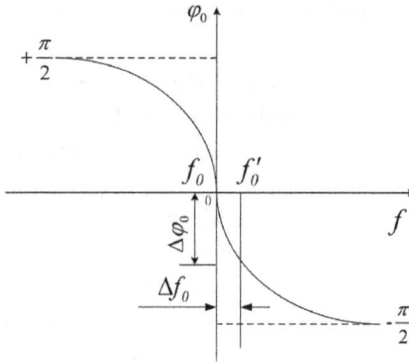

Fig. 3.6: Phase stabilization of a resonant circuit.

We can assume a sudden change of frequency from f_0 to $f_{0'}$ ($\Delta f_0 = f_0{'} - f_0 > 0$) is triggered as shown in Fig. 3.6. The instantaneous phase $\varphi(t) = \varphi_0 + \int f_0{'}dt$ suddenly becomes larger because of frequency instability. According to the phase–frequency curve with negative slope, phase difference $-\Delta\varphi_0$ is then introduced compensating the phase increase $\varphi(t) = \varphi_0 + \int f_0{'}dt - \Delta\varphi_0$. The compensation of phase will not stop until the frequency gets back to f_0 again. Therefore, a steeper phase–frequency curve which introduces larger $-\Delta\varphi_0$ helps the system restore balance faster.

We can conclude that the phase stability condition is

$$\left.\frac{d\varphi}{df}\right|_{f=f_0} < 0 \tag{3.7}$$

Frequency stability condition can be analyzed in the similar way by exchanging the axis of frequency and phase curve.

3.2.2.4 Discussion on oscillation conditions

(1) None of the earlier six conditions, eqs. (3.4) to (3.7), can be dispensed for oscillators. Among the six conditions, stability conditions are the most implicit and rely on circuit structure.

(2) If the circuit is designed properly, it should automatically balance once it successfully starts.

(3) The above-mentioned six oscillation conditions can be summed up as three criterions.

   ① Positive feedback.

   ② $AF > 1$.

   ③ The phase–frequency characteristic of the frequency-selective filter is with negative slope.

(4) Analysis of the oscillator can be performed in two steps.
   ① Qualitative analysis: Sanity check of the circuit structure. For example, it should be checked that whether the oscillator contains positive feedback or whether the phase-frequency property is with negative slope.
   ② Quantitative analysis: verification of the amplitude startup condition: $AF > 1$. Since the transistor is in the linear amplifying state when the oscillator starts working and the startup signal is very weak, the circuit can be analyzed by the method of micro variable equivalent circuit.

3.3 Feedback-type *LC* oscillators

In an *LC* oscillator, the filter is a tuned or resonant circuit(often called a tank circuit) consisting of both inductors (*L*) and capacitors (*C*). The charge flows back and forth between the capacitor and the inductor, so that the tuning circuit can store the oscillating electrical energy at the resonant frequency. There is a small loss in the tank circuit, but the amplifier compensates for these losses and provides power to the output signal [11].

Feedback-type *LC* oscillators are the most frequently used *LC* oscillators. The two categories of feedback- type *LC* oscillators are capacitance feedback and inductance feedback, such as Colpitts oscillator, Hartley oscillator, Clapp oscillator and Seiler oscillator.

The feedback *LC* oscillators are designed for sinusoidal oscillations and often used at radio frequencies when a tunable frequency source is necessary, such as in function generators, tunable radio transmitters and local oscillators in radio receivers.

3.3.1 Principle of feedback-type *LC* oscillators composition

There is a simple rule for all feedback-type *LC* oscillators: The component between emitter and base, and the component between emitter and collector should be either both capacitive or both inductive; the component between base and collector is opposite in reactance polarity to the previous two. This rule is in fact summed up by the phase balance condition of oscillators.

We can explain this problem in Fig. 3.7. The following two figures are the basic composition of capacitance feedback and inductance feedback *LC* oscillators. Figure 3.7(a) is known as Colpitts circuit and Fig. 3.7(b) is Hartley circuit.

Both of them use an *LC* circuit in the feedback loop to provide necessary phase shift for positive feedback and to act as a frequency-selective filter that passes only the desired frequency of oscillation. The basic configuration of two circuits is similar but the feedback voltage is obtained by capacitive divider (Colpitts) and inductive divider (Hartley).

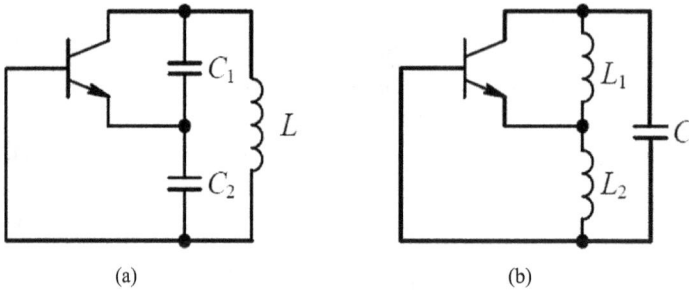

Fig. 3.7: Two basic types of feedback *LC* oscillator: (a) Colpitts circuit and (b) Hartley circuit.

In the following analysis of phase condition, we use Fig. 3.8 for a generalized case. For simplification loop loss such as the input and output impedance of transistor is ignored and circuit quality factor is assumed to be high enough.

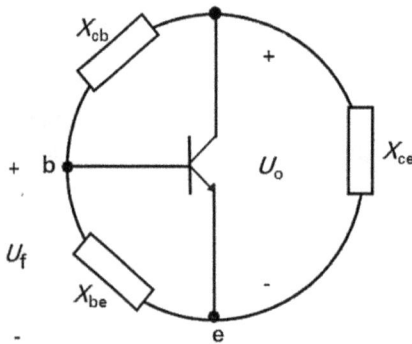

Fig. 3.8: The feedback *LC* oscillator composition.

When the circuit is in resonance, it is purely resistive

$$X_{be} + X_{ce} + X_{cb} = 0 \tag{3.8}$$

The feedback voltage U_f determined by the reactance divider is

$$U_f = \frac{jX_{be}}{j(X_{be} + X_{cb})} \cdot U_o = -\frac{X_{be}}{X_{ce}} U_o \tag{3.9}$$

As for transistors, the output voltage U_o and the input voltage U_i are out of phase. The phase balance condition requires that U_i and U_f are in phase. So U_f and U_o should be out of phase, that is, with 180° phase difference.

Therefore, the criteria of feedback *LC* oscillator composition are
(1) Reactance X_{be} and X_{ce} are the same type (both capacitive or both inductive).
(2) Reactance X_{cb} is different from X_{be} or X_{ce}.

3.3.2 Colpitts circuit

The most basic capacitance feedback oscillator is Colpitts circuit shown in Fig. 3.9. Colpitts oscillator is designed for generation of sinusoidal wave in the range of 20 kHz–30 MHz. It is often used in radio receivers as a local oscillator.

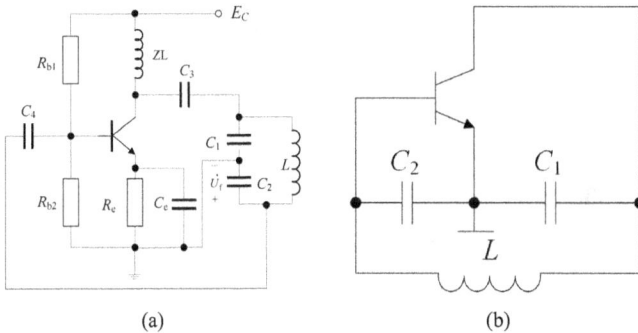

Fig. 3.9: Colpitts oscillator, (a) schematic circuit, (b) AC equivalent circuit.

In Fig. 3.9(a), a Colpitts oscillator consists of an R–C coupled amplifier using an NPN transistor in common emitter configuration. R_{b1} and R_{b2} are two biasing resistors for the transistor. Resistor R_e is connected in the circuit to stabilize the circuit against temperature variations. Capacitor C_e is a bypass capacitor providing a low reactive path to AC signals. C_3 and C_4 are coupling capacitors which provide AC path from collector or base to the LC network. ZL is the RF choke which prevents AC signals from affecting the DC power supply. The LC network (tank circuit) consists of an inductor L and two capacitors C_1 and C_2 (in series). The feedback signal is obtained from capacitor C_2 and is sent back to the base electrode.

The AC equivalent circuit of Colpitts oscillator is shown in Fig. 3.9(b). C_1, C_2 and L form the resonance circuit with the relationship of

$$X_{C1} + X_{C2} = X_L \tag{3.10}$$

The circuit adopts common emitter configuration and the output signal is 180° inverse phase with respect to the input signal. The additional required 180° phase shift comes from the LC network. A fraction of the output signal is fed back to the base of transistor.

When the phase balance condition is met and quality factor Q of the LC resonant circuit is high enough, the oscillating frequency approximately equals the LC resonant circuit frequency:

$$f_0 \approx \frac{1}{2\pi\sqrt{LC}} \tag{3.11}$$

where C is the capacitance of C_1 and C_2 connected in series:

$$C = \frac{C_1 C_2}{C_1 + C_2} \tag{3.12}$$

Feedback coefficient is determined by capacitance C_1 and C_2. The two capacitors are usually "combined" together to supply a constant amount of feedback so that when one is automatically adjusted and the other is automatically followed. We can see that the voltage across C_1 is the same as the oscillator's output voltage, while the voltage across C_2 is the feedback voltage of oscillator. For the Colpitts circuit shown in Fig. 3.9, the feedback coefficient is therefore simply calculated as

$$F \approx \frac{C_1}{C_2} \tag{3.13}$$

By adjusting the capacitance of C_1 and C_2, the amplitude of feedback voltage can be tuned. Too large feedback may cause distortion of the output signal, while too small feedback may not allow the circuit to oscillate.

To sum up, advantages of a Colpitts circuit include [12] the following:
(1) The output waveform is good with little distortion because the feedback voltage is obtained from the capacitor.
(2) High-frequency stability (10^{-3}) can be achieved. Proper increase of circuit capacitance can reduce the influence of unstable factors on oscillation frequency.
(3) High oscillation frequency can be realized. The input capacitance or the output capacitance of transistor can be used as resonant capacitance, so the typical frequency is in the dozens of MHz to hundreds of MHz range.

There are also disadvantages of Colpitts circuits:
(1) Both oscillation frequency and feedback coefficient depend on C_1 and C_2. When adjusting C_1 or C_2 to change the frequency, the feedback coefficient also changes, which affects the amplitude of oscillation signal and even fails the oscillation condition.
(2) The capacitance of transistor contributes to the output frequency but transistor properties are easily influenced by temperature, biasing condition, power supply and so on. Instability factors which may diminish the frequency stability of oscillators are introduced.

3.3.3 Hartley circuit

Another important and basic feedback-type *LC* oscillator is Hartley circuit shown in Fig. 3.10. Two inductors L_1 and L_2 are used across a single common capacitor C, jointly forming the resonant circuit [13].

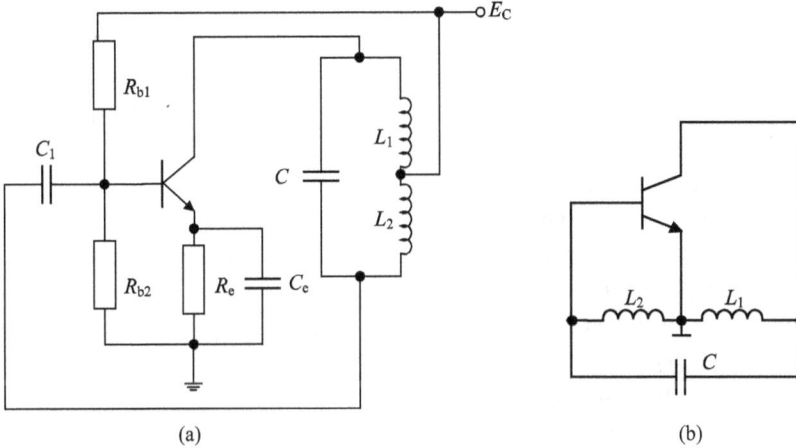

Fig. 3.10: Hartley oscillator: (a) basic circuit composition and (b) AC equivalent circuit.

In Fig. 3.10(a), Hartley oscillator consists of an R–C coupled amplifier using an NPN transistor in common emitter configuration. Again, here R_{b1} and R_{b2} are biasing resistors, and R_e is responsible for stabilizing the circuit against temperature variations. C_e and C_1 are bypath and coupling capacitors, respectively. The inductor is tapped in middle to provide feedback signal (voltage on inductor L_2) back into the base electrode.

When DC supply E_C is on, the collector current begins to rise and charge the capacitor C. When the capacitor is fully charged, it is discharged to coils L_1 and L_2 to form damping harmonic oscillations in the tank circuit. The resulting voltage span inductor L_1 is the output signal, and the voltage across L_2 is sent back to the b–e junction of transistor as a feedback signal. Coil L_1 and coil L_2 are coupled by mutual inductance M. DC energy is converted into sustainable oscillation outputs and circuit loss.

Because the tap position is AC grounding, the phase difference between L_1 and L_2 is always 180°. In addition, the phase difference 180° between the input and the output voltage of transistor is also introduced. Therefore, the total phase shift becomes 360° (or zero), so that the feedback is positive or regenerated, which is necessary to oscillate. Therefore, it is possible to oscillate continuously without damping. The resonant circuit (L_1, L_2 and C) determines the frequency of oscillator which is calculated as

$$f_0 \approx \frac{1}{2\pi\sqrt{LC}} \qquad (3.14)$$

where L is the total loop inductance that is given by L_1, L_2 and mutual inductance M

$$L = L_1 + L_2 + 2M \tag{3.15}$$

In this case, the amplitude of feedback signal depends on the feedback coefficient which is related to L_1 and L_2 and is given by

$$F \approx \frac{L_2 + M}{L_1 + M} \tag{3.16}$$

By tuning the inductance of L_1 and L_2, we can adjust the F value. Similarly, either too large or too small feedback will not be a good choice for the oscillator circuit. We need to balance the distortion possibility and feasibility of oscillation and then determine the optimal range for F.

Advantages of Hartley oscillator include the following:
(1) It is easy to start oscillating since there is strong mutual inductance between L_1 and L_2 ensuring strong feedback.
(2) Frequency is easily adjustable as by tuning capacitance of C.
(3) Feedback coefficient F is independent of frequency adjustment since F depends mainly on L and frequency is tuned by C.

Disadvantages of Hartley oscillator include the following:
(1) The output waveform is poor because the feedback voltage is obtained from the inductor, which inevitably contains multiple harmonic components.
(2) High oscillation frequency is hardly achievable. Because when frequency is getting very high, the effect of interelectrode capacitance increases, and the property of circuit reactance can change and fail the necessary phase condition.
(3) Transistor instability can still affect oscillation frequency just like Colpitts oscillator.

Example 3.1: The AC equivalent circuit of an LC oscillator is shown in Fig. 3.11. The resonant frequencies of the three parallel LC loops are $f_1 = \frac{1}{2\pi\sqrt{L_1 C_1}}$, $f_2 = \frac{1}{2\pi\sqrt{L_2 C_2}}$ and $f_3 = \frac{1}{2\pi\sqrt{L_3 C_3}}$. Find out the quantitative relation between f_1, f_2 and f_3 so that the oscillator meets the phase startup condition.

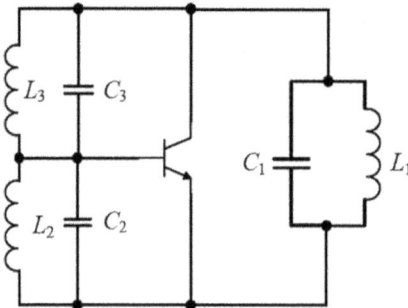

Fig. 3.11: Circuit of Example 3.1.

Solution: There are two possibilities that the oscillator meets the phase condition.

(1) If it is the capacitance feedback type: At the oscillation frequency f_{osc}, L_1C_1 loop and L_2C_2 loop are capacitive, and the L_3C_3 loop is inductive. That is to say f_{osc} is greater than f_1 and f_2 but less than f_3. So f_1, f_2 and f_3 should meet the quantitative relation of $f_1, f_2 < f_{osc} < f_3$.

(2) If it is the inductance feedback type: At the oscillation frequency f_{osc}, L_1C_1 loop and L_2C_2 loop are inductive, and the L_3C_3 loop is capacitive. That is to say f_{osc} is less than f_1 and f_2 but larger than f_3. So f_1, f_2 and f_3 should meet the quantitative relation of $f_1, f_2 > f_{osc} > f_3$.

3.4 Improved capacitance feedback *LC* oscillators

As discussed in the disadvantage parts, the oscillation frequency of earlier two oscillators are not only determined by *LC* resonant circuit parameters, but also related to the input capacitance C_i and the output capacitance C_o of transistor. When the environment is changed or the transistor is replaced, frequency stability or consistency cannot be guaranteed. As shown in Fig. 3.12, the transistor capacitances C_o and C_i are in fact in parallel with the capacitance C_1 and C_2.

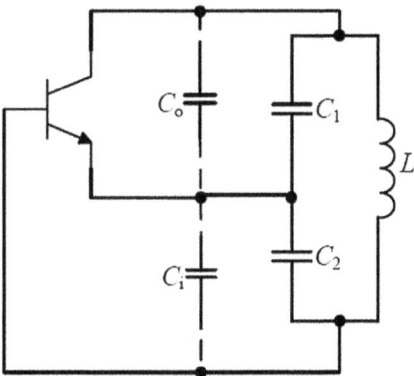

Fig. 3.12: Equivalent circuit of an *LC* oscillator when the effect of C_i and C_o is included.

Neglecting other possible influences, the oscillation frequency is modified to be

$$\omega_0 \approx \frac{1}{\sqrt{L\frac{(C_1+C_o)(C_2+C_i)}{C_1+C_2+C_o+C_i}}} \tag{3.17}$$

How to reduce the influence of C_i and C_o and improve the frequency stability? It seems that increasing the value of loop capacitance C_1 and C_2 can help reduce the influence of C_i and C_o on the frequency. But this method is only applicable when frequency is not too high (large C_1 and C_2 limit frequency). When frequency is high, the increase of C_1 and C_2 requires the reduction of inductance L to maintain high oscillation frequency. This leads to the decrease of Q factor and the oscillation amplitude and even makes the oscillation to stop. Further improvements are needed in order to weaken the negative effects of transistor.

3.4.1 Clapp circuit

The first improved capacitance feedback LC oscillator which is called Clapp circuit (Fig. 3.13) is introduced. Clapp oscillator is a variation of Colpitts oscillator. The main difference lies in an additional variable capacitor C_3 in series with inductor L. The inductor in Colpitts oscillator is replaced by the L–C_3 combination in Clapp oscillator.

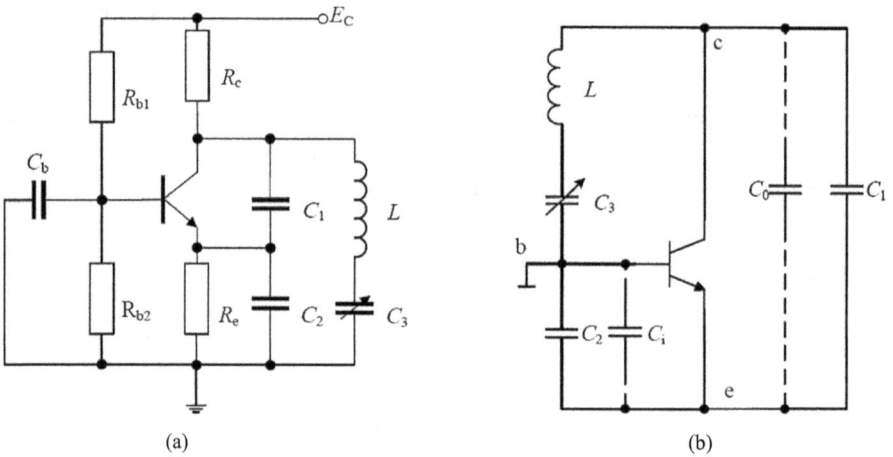

Fig. 3.13: Clapp oscillator: (a) basic circuit composition and (b) AC equivalent circuit.

The circuit principle is basically the same as Colpitts oscillator. A properly selected C_3 capacitance can improve frequency stability by screening the effect of transistor parameters. Usually C_3 is chosen to be much smaller than C_1 and C_2 and dominates oscillation frequency.

It can be seen from the effective total capacitance C_Σ which is calculated as follows:

$$C_\Sigma = \frac{C_1'C_2'C_3}{C_1'C_2' + C_1'C_3 + C_2'C_3} \approx C_3 \tag{3.18}$$

where $C_1' = C_1 + C_o$ and $C_2' = C_2 + C_i$. Therefore, the oscillation frequency is approximately

$$f_0 \approx \frac{1}{2\pi\sqrt{LC_3}} \tag{3.19}$$

According to eq. (3.19), we can see that the oscillation frequency is almost unrelated to C_1, C_2, C_o or C_i. Hence, the property variation of transistor has little influence on frequency, and frequency stability is improved. This makes Clapp oscillator superior to Colpitts oscillator.

As discussed, small capacitance C_3 is preferred for the consideration of frequency stability. However, it is not always the smaller the better. In order to explain this point well, we start equivalent output resistance of circuit (Fig. 3.14).

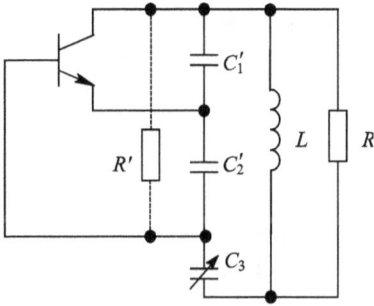

Fig. 3.14: Resonant resistance R is converted to the output resistance R' of oscillator.

First, we convert the resonant resistance R of LC network to the equivalent output resistance R' of oscillator. This is meaningful because output resistance decides the amplitude of output voltage and thus the gain, which is an important performance index of an oscillator circuit. According to the calculation principle of partial access in the first chapter, R' can be written as

$$R' = n^2 R \tag{3.20}$$

where the partial access coefficient n, also called the voltage division ratio, is given by (let $C_{1'} = C_1 + C_o$, $C_{2'} = C_2 + C_i$)

$$n = C_3 \bigg/ \left(C_3 + \frac{C_1' C_2'}{C_1' + C_2'} \right) = \frac{1}{1 + \frac{C_1' C_2'}{(C_1' + C_2') \cdot C_3}} \tag{3.21}$$

And the resonant resistance R is known as

$$R = Q\omega_0 L \tag{3.22}$$

Based on the earlier expressions, when capacitance C_3 is too small, n and R' are both small. The voltage gain of amplifier and the output signal amplitude of oscillator will be greatly reduced, which may be the cause for the stop of oscillation.

In conclusion, reducing capacitance C_3 to improve the oscillation frequency stability is at the cost of lower voltage gain.

As for Clapp oscillator, there are several defects or limitations listed as follows:

(1) When capacitance C_3 is reduced to improve frequency stability, the output amplitude will significantly decrease; when C_3 is reduced to a certain extent, the oscillator may stop oscillating, thus the frequency stability of oscillation is limited.

(2) When used as a tunable oscillator, the oscillation amplitude is still dependent with frequency. The output amplitude is not constant in the band range, and the frequency coverage coefficient (the ratio of high frequency and low frequency) is not large, 1.2–1.3.

Based on the limitations of Clapp oscillator, a further improved circuit is proposed in the next section.

3.4.2 Seiler circuit

To solve the problems existed in Clapp oscillator, another capacitance feedback type circuit, Seiler oscillator, is proposed. Compared with Clapp oscillator, in Seiler oscillator an additional variable capacitor C_4 is in parallel with L for frequency tuning (Fig. 3.15).

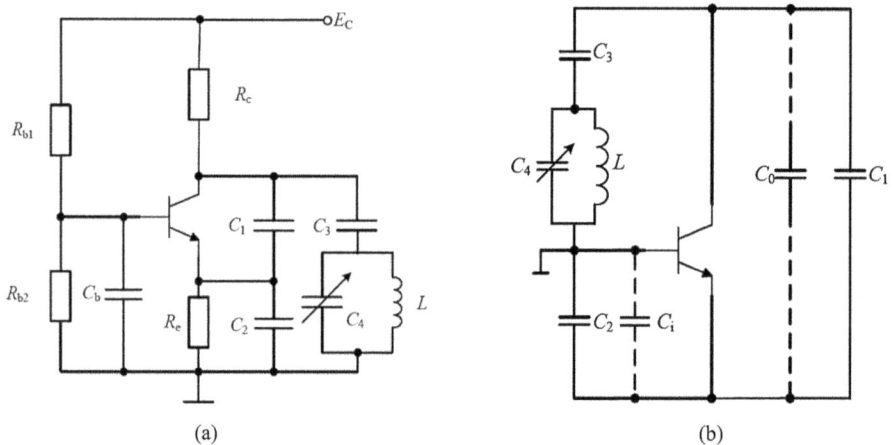

Fig. 3.15: Seiler oscillator (a) basic circuit composition and (b) AC equivalent circuit.

The circuit principle is very similar to Clapp oscillator. In Seiler oscillator, capacitance C_3 and C_4 determines the oscillation frequency; C_3 is chosen to be much smaller than C_1 and C_2 to screen the effect of the transistor and ensure frequency stability.

The effective total capacitance C_Σ is calculated as

$$C_\Sigma = C_4 + \frac{1}{1/C_1 + 1/C_2 + 1/C_3} \approx C_4 + C_3 \tag{3.23}$$

Consequently, the oscillation frequency is approximately

$$f_0 \approx \frac{1}{2\pi\sqrt{L(C_3 + C_4)}} \tag{3.24}$$

In Colpitts oscillator, the resonant frequency is affected by the transistor and stray capacitance because C_1 and C_2 are shunted by the transistor and stray capacitance, and so their values are altered more or less. But in Seiler oscillator, the transistor and stray capacitance has little effect on capacitance C_3 and C_4, so the oscillation frequency is more stable and accurate.

Let us take a look at the equivalent output resistance of Seiler circuit (Fig. 3.16) and see how we should determine C_4 and C_3 values in order to balance the output amplitude (or gain) and frequency stability in this scenario.

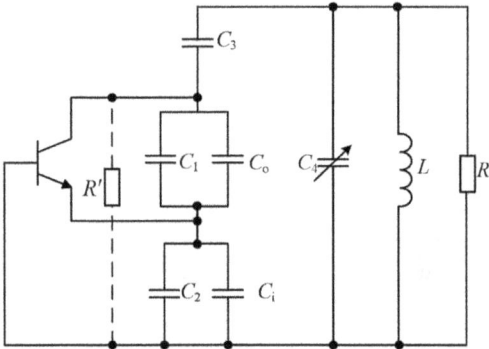

Fig. 3.16: Resonant resistance R is converted to the output resistance R' of oscillator.

The resonant resistance R is converted to the output of transistor as R' by the same technique:

$$R' = n^2 R \tag{3.25}$$

where the partial access coefficient n is (let $C_{1'} = C_1 + C_o$, $C_{2'} = C_2 + C_i$)

$$n = C_3 \left/ \left(C_3 + \frac{C_1' C_2'}{C_1' + C_2'} \right) \right. \tag{3.26}$$

And the resonant resistance R is

$$R = Q\omega_0 L \tag{3.27}$$

Access coefficient n of Seiler circuit shown in eq. (3.26) looks the same as Clapp circuit in eq. (3.21). However, here n is unrelated to capacitance C_4. So, when adjusting the capacitance C_4 to change oscillation frequency, either n or R' will not be affected, maintaining constant voltage gain and stable output signal amplitude.

Compared with Clapp oscillator, Seiler oscillator breaks the unwanted connection between gain and frequency, which realizes relatively stable output amplitude and the wider frequency tuning range. The coverage coefficient of frequency is high, generally 1.6–1.8. In practice, Seiler oscillators have been widely used in broadband systems.

3.5 Frequency stability of oscillators

One of the most important performance indicators of oscillators is frequency stability, in other words, the ability to provide constant frequency output under complex conditions. Unstable frequency in a communication system is responsible for signal loss. Frequency instability or inconsistency of measuring apparatus results in errors. The carrier frequency offset causes distortion of the speech over telephone.

3.5.1 Qualitative analysis of frequency stability

There are two ways to quantitatively describe frequency stability.

3.5.1.1 Absolute frequency stability
Δf is defined as the absolute deviation of f away from a standard frequency f_0 under certain conditions:

$$\Delta f = f - f_0 \tag{3.28}$$

3.5.1.2 Relative frequency stability
It is defined as the ratio of absolute frequency stability Δf and standard frequency f_0 under certain conditions:

$$\frac{\Delta f}{f_0} = \frac{f - f_0}{f_0} \tag{3.29}$$

Relative frequency stability is more commonly used. Small $\Delta f/f_0$ indicates higher frequency stability. For instance, if the reference frequency of an oscillator is 1 MHz and the actual working frequency is 0.99999 MHz, then the relative frequency stability is $\Delta f/f_0 = 10\ \text{Hz}/1\ \text{MHz} = 10^{-5}$. Certain conditions mentioned earlier refer to the concerned time scale, temperature range or voltage variation conditions while measuring frequency stability.

When talking about different time scales, here is a rough classification:

(1) Short-term frequency stability: typically, within 1 h or less. It is generally used to evaluate frequency stability of measuring instruments and communication equipments.

(2) Medium-term frequency stability: within one day. Medium-term frequency stability of the medium wave radio transmitter is 2×10^{-5}/day; television transmitting station's is 5×10^{-7}/day; LC oscillator's generally is 10^{-4}–10^{-3}/day; Clapp's oscillator and Seiler's oscillator are 10^{-5}–10^{-4}/day.

(3) Long-term frequency stability: within months, one year or even longer time periods.

3.5.2 Factors causing frequency instability

The frequency of oscillator depends on circuit parameters as well as transistor parameters. These parameters cannot be strictly constant all the time under complicated conditions, so the oscillation frequency cannot be absolutely stable.

3.5.2.1 Unstable parameters of LC circuit

Temperature change is one of the main factors that disturb LC parameters. Temperature variation may result in geometry changes for capacitors or inductance coils, thereby changing the values of L and C. Generally, inductance is with a positive temperature coefficient, that is, inductance increases with the temperature rise. Temperature coefficient for capacitance could be either positive or negative depending on dielectric materials and structures.

Besides, mechanical vibration also causes device deformation and L and C value changes, thus results in the instability of oscillation frequency.

3.5.2.2 Unstable parameters of transistor

Transistors are influenced by temperature and DC power supply largely. When the quiescent point and junction capacitance are shifted by the mentioned factors the oscillation frequency also deviates from expectation.

3.5.3 Measures for stabilizing frequency

Based on the above discussion, frequency stability can be improved from two aspects: first, reduce the impact from external or environmental factors; second, improve the standardization of resonant circuit components as well as the antiinterference ability of circuit.

3.5.3.1 Reducing the impact from external factors

External factors that can affect frequency stability of oscillators typically include and not limited to temperature, the load, DC power supply, ambient electromagnetic field, humidity and mechanical vibration. Among those factors temperature plays the most important role on the oscillator performance.

(1) Reducing the effect of temperature variation. There are multiple methods to reduce temperature variation. One way is to use thermostatic bath for very high precision purposes. Also, inductor and capacitor with low temperature coefficient are more preferred. For example, the inductance coil can be used for the high-frequency magnetic skeleton, which has lower temperature coefficient and loss. For air variable capacitors, copper is a better material than aluminum when used as bracket because of copper's lower coefficient of thermal expansion. Among all fixed-value capacitors, Mica capacitor is the one with lowest temperature coefficient and most reliable performance.

(2) Stabilizing power supply voltage. Power supply fluctuations will change the DC operating point of transistor and further change the transistor parameters. In order to reduce the effect, we can choose well-regulated power supply and a stable operating point.

(3) Reducing the effect of load. Output signal drives the load, but load may need to be replaced in some circumstances. The change of load resistance or reactance inevitably makes the oscillation frequency inconsistent. In order to reduce this impact, a buffer can be applied between the oscillator and the load, which is composed by the emitter follower with large input resistance, thereby weakening the impact of load on circuit.

(4) Shielding the circuit and keeping it away from heat source or sink. Shielding the circuit can minimize the interference of electromagnetic field. Moreover, it is wise not to place the circuit near any heat source or sink. To be more specific, we should keep the oscillator away from ovens, high-power transistors, power transformer, voltage regulators, rectifiers and lamps.

3.5.3.2 Improving standardization of resonant circuit

The standardization of resonant circuit refers to the ability to maintain the resonant frequency of circuit even if the test environment changes.

(1) Using inductors and capacitors with high stability. In order to ensure the stability of circuit's resonance frequency, it is better to use inductors and capacitors with high-quality materials, reasonable structure and advanced technology.

(2) Loose coupling between the transistor and the loop. Taking Clapp and Seiler oscillators as examples, those inductors and capacitors which dominate the oscillation frequency are separated from the input and the output capacitance (C_i and C_o) of transistor, forming weak coupling between the transistor and the resonant loop.

(3) Improving Q of resonant circuit. We know that phase change is related to frequency change. According to the phase stability condition, the phase–frequency characteristic curve of resonant circuit should be with negative slope. The relationship between phase φ and Q of the resonant circuit is

$$\varphi = - \arctan 2Q\left(\frac{\Delta\omega}{\omega_0}\right) \tag{3.30}$$

The slope of phase–frequency characteristic is directly proportional to Q value. That is to say, with greater Q the phase–frequency curve appears steeper. According to our previous analysis, a steeper curve benefits the circuit by compensating the phase error faster and stabilizes in a shorter period of time. Therefore, with higher Q factor of resonant circuit, frequency stability is anticipated to be improved.

3.6 Quartz crystal oscillators

With the development of modern electronics and communication, the increasing demand for high-frequency stability devices drive novel discoveries on materials and processing technology. Though many stabilizing measures can be applied on LC oscillators, the typical frequency stability is still in the order of 10^{-5}. LC circuit mainly suffers from device-limited Q-factor of 200 and less. In searching for higher Q materials, people found quartz crystal an attractive candidate for the application of high precision oscillators. Quartz crystal, or crystal for short, is a material with piezoelectricity and inverse-piezoelectricity. When the crystal is connected into a circuit by electrodes, its mechanic vibration induces electric oscillation with the same frequency. Crystals have surprisingly high Q factors (in excess of 10^4 can be achieved), very low temperature coefficient and other excellent physical and chemical properties. Therefore, the frequency stability can be up to 10^{-10}–10^{-11} for crystal oscillators which is a better way than LC oscillators. The advantages of quartz crystal motivate wide range of applications to thrive. Next, we will start by introducing the basic electrical characteristics of quartz crystal.

3.6.1 Equivalent circuit and electrical properties of quartz crystal

Quartz crystal is a crystal form of silica with the chemical composition of SiO_2. Quartz crystal used in a crystal oscillator is a very small and thin cut quartz piece or chip with its upper and bottom surfaces metalized as electrodes for electrical connections.

3.6.1.1 Piezoelectricity and the equivalent circuit of quartz crystal
Quartz crystal is both piezoelectric and inverse-piezoelectric. When voltage is applied to quartz crystal along a certain axis, lattice structure deforms accordingly

and this property is known as piezoelectricity. Inversely, mechanical force exerted on certain surfaces causes the crystal to induce electric response of voltage.

Such properties can be used in transducers as they convert energy from one form to another (electrical to mechanical or mechanical to electrical). The oscillation produced by quartz crystals can perfectly replace LC tank circuit in previously introduced oscillators. The frequency of crystal oscillation is mainly determined by its geometry. Once shaped and secured in place, the crystal can only operate at one resonant frequency.

Figure 3.17(a) shows the circuit symbol for a quartz crystal. The electrical effect of a mechanically vibrating crystal can be represented by an equivalent circuit as shown in Fig. 3.17(b). In this circuit, a large static C_0 represents the distributive capacitance due to the coated electrodes. The parallel RLC branch represents the mechanical vibration property of the crystal, in which L_q is the large inductance corresponding to the kinetic energy of vibration, C_q is the small capacitance corresponding to the elastic potential energy and r_q is the low dynamic resistance corresponding to the friction or energy loss in the form of heat.

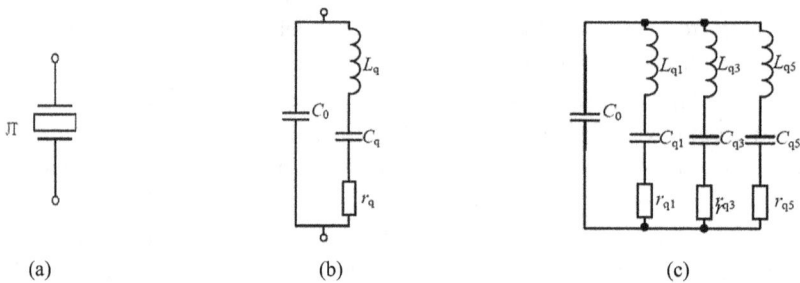

Fig. 3.17: Quartz crystal, (a) Circuit symbol, (b) fundamental equivalent circuit, (c) complete equivalent circuit.

Crystals can also be made to resonate at higher odd multiples of their basic resonant frequency, with the higher odd harmonics also known as overtones. Overtones can then be filtered out using a conventional LC circuit. So, the complete equivalent circuit of quartz crystal is shown in Fig. 3.17(c). By using overtone technique, wider operational range of crystal frequencies can be achieved.

The most impressive feature of quartz crystal is that the inductance L_q is usually measured to be very large but capacitance C_q and resistance r_q are measured very small. The resulting Q-factor is as high as 10^4–10^6, so the oscillation frequency stability of quartz crystal resonator is significantly higher than conventional resonators.

3.6.1.2 Impedance characteristic of quartz crystal

It can be seen from equivalent circuit shown in Fig. 3.17(b) that quartz resonator has two resonant frequencies: the series resonant frequency f_s and the parallel resonant frequency f_p.

(1) Series resonant frequency f_s. In the equivalent circuit, L_q and C_q constitute the series resonant circuit and the series resonant frequency f_s is

$$f_s = \frac{1}{2\pi\sqrt{L_q C_q}} \tag{3.31}$$

(2) Parallel resonant frequency f_p. In the equivalent circuit, L_q, C_q and C_0 constitute the parallel resonant circuit together and the parallel resonant frequency f_p is

$$f_p = \frac{1}{2\pi\sqrt{L_q \frac{C_0 C_q}{C_0 + C_q}}} \tag{3.32}$$

f_p can be expressed in the form of f_s as

$$f_p = \frac{1}{2\pi\sqrt{L_q C_q}}\sqrt{L_q \frac{C_0 C_q}{C_0 + C_q}} = f_s\sqrt{1 + \frac{C_q}{C_0}} \tag{3.33}$$

From this expression, we observe that the two resonant frequencies are very close since $C_0 \gg C_q$.

(3) The reactance characteristic of quartz crystal. The equivalent resistance r_q is very small in the *RLC* branch. For simplicity we ignore the effect of r_q in the circuit model (Fig. 3.18(a)). The equivalent reactance X with respect to frequency f is plotted in Fig. 3.18(b) with detailed deductions followed.

According to the circuit model, it is not difficult to begin writing and vary the expression of the equivalent impedance Z as

$$Z = \frac{j[\omega L_q - 1/(\omega C_q)](-j\omega C_0)}{j[\omega L_q - 1/(\omega C_q) - 1/\omega C_0]}$$

$$= -j\frac{1}{\omega C_0}\frac{\omega L_q[1 - 1/(\omega^2 L_q C_q)]}{\omega L_q\left[1 - 1/\left(\omega^2 L_q \frac{C_q C_0}{C_q + C_0}\right)\right]}$$

$$= -j\frac{1}{\omega C_0}\left(1 - \frac{\omega_s^2}{\omega^2}\right)\Big/\left(1 - \frac{\omega_p^2}{\omega^2}\right) \tag{3.34}$$

where ω_s and ω_p are series resonant frequency and parallel resonant frequency, respectively. The expression of Z is pure reactance since resistance r_q is ignored as a premise. Reactance X (same as Z) is then plotted in Fig. 3.18(b).

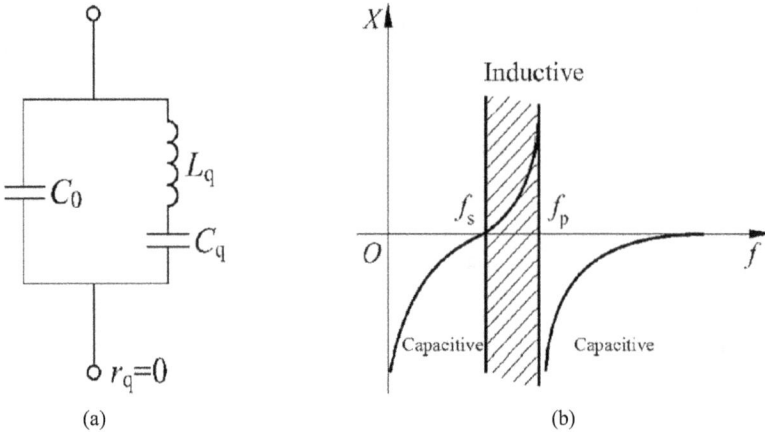

Fig. 3.18: Quartz crystal's characteristic, (a) equivalent circuit, (b) reactance.

When $\omega = \omega_s$, resonance occurs in the series branch and the total impedance $X = 0$ (short circuit). When $\omega = \omega_p$, resonance occurs in the parallel loop and $X \to \infty$. When $\omega < \omega_s$ or $\omega > \omega_p$, X is negative, so the circuit is capacitive. When $\omega_s < \omega < \omega_p$, X is positive and the circuit is inductive.

The separation between ω_s and ω_p being very small, the inductive region (in shadow) is quite narrow resulting in the steep reactance curve between ω_s and ω_p. In fact, this extremely narrow inductive region area is the most popular frequency band of quartz crystals which are applied as highly stable inductors. The very large slope contributes to high Q factor hence excellent frequency stability.

3.6.1.3 Reasons of high-frequency stability of quartz crystal
(1) Its temperature coefficient is very small. With the constant temperature equipment, the higher stability of frequency is more guaranteed.
(2) High Q-factor property in the frequency range between f_s and f_p: phase–frequency characteristic curve changes very fast, which is good for frequency stabilization.
(3) $C_q \ll C_0$: the oscillation frequency is basically decided by L_q and C_q and the external circuit has little impact on the oscillation frequency. This can be further explained in two aspects:
  ① Assume a distribution capacitor C_n is added in parallel with C_0 (Fig. 3.19):

Because $C_0 \gg C_q$ and $(C_0 + C_n) \gg C_q$, the oscillation frequency is modified as

$$\omega = \frac{1}{\sqrt{L_q \left[\dfrac{C_q(C_0 + C_n)}{C_q + C_0 + C_n}\right]}} \approx \frac{1}{\sqrt{L_q C_q}} \tag{3.35}$$

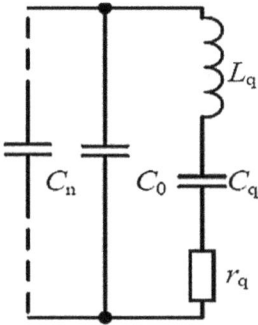

Fig. 3.19: A distributed capacitor C_n is in parallel with C_0.

Clearly, the oscillation frequency is nearly unaffected after the addition of an external C_n, which is dominated by parameters L_q and C_q.

② Assume an external resistor R is added in parallel with C_0 (Fig. 3.20):

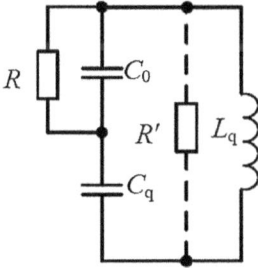

Fig. 3.20: An external resistor R is added to circuit and converted.

The resistance R is converted to the two ends of resonant loop as R'

$$R' = \left(\frac{C_q + C_0}{C_q}\right)^2 R \approx \left(\frac{C_0}{C_q}\right)^2 R \tag{3.36}$$

Since $C_0 \gg C_q$, then the converted R' is far larger than the actual value R. The shunting effect on inductance L_q of external resistance is very small, and the quartz resonator can still keep high Q.

3.6.2 Quartz crystal oscillator circuits

Crystal oscillators are widely used because of its excellent frequency stability (10^{-10}–10^{-11}), and they are employed in most of the radio transmitters and computers or quartz clocks to generate the clock signals. Many crystal oscillators share similar circuit structure as LC oscillators, with the crystal replacing the frequency selection part. Pierce oscillator circuit is a typical example. Crystal oscillators can provide sine wave output over a wide frequency range from several MHz up to several hundred of MHz.

There are various kinds of crystal oscillators. Based on the function in circuit, crystal oscillators usually fall into one of the two categories: "parallel mode," in which crystal basically works as an inductor or "series mode" in which the crystal acts as a highly selective low impedance device at around f_s in the positive feedback path.

3.6.2.1 Crystal oscillators in parallel mode

3.6.2.1.1 Pierce oscillator

Pierce oscillator is a good example of crystal oscillator in parallel mode. From the schematic circuit of Pierce oscillator shown in Fig. 3.21(a), the circuit is similar to the capacitance feedback LC oscillator. Crystal is set to operate at a frequency between f_s and f_p, behaving as an inductor, which can be compared with the inductor L in Colpitts oscillator.

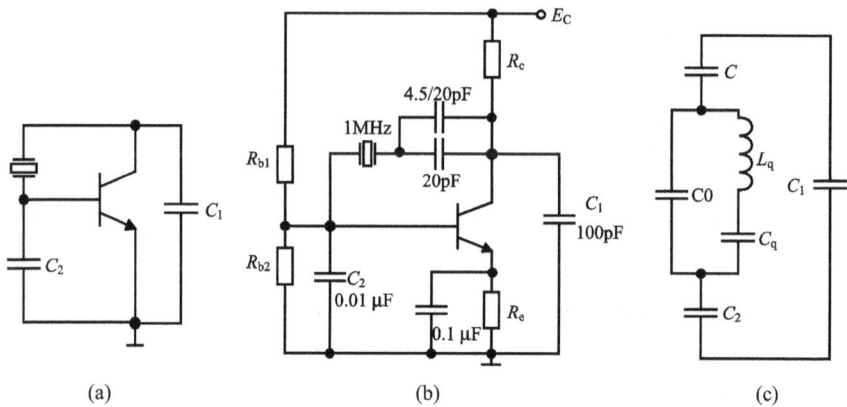

Fig. 3.21: Pierce oscillator: (a) schematic circuit; (b) actual circuit; and (c) equivalent resonant circuit.

Figure 3.21(b) presents the actual circuit of Pierce oscillator. Crystal together with two small capacitors 4.5/20 and 20 pF, as well as C_1 and C_2 form the resonant loop, with the designed resonant frequency between f_s and f_p.

(1) Oscillation frequency f_0. The equivalent resonant circuit of Pierce oscillator is shown in Fig. 3.21(c), where capacitance C stands for the combined tunable effect of 4.5/20 and 20 pF capacitors. The total capacitance C_Σ responsible for oscillation frequency can be calculated as

$$\frac{1}{C_\Sigma} = \frac{1}{C_q} + \frac{1}{C_0 + \frac{1}{\frac{1}{C} + \frac{1}{C_1} + \frac{1}{C_2}}} \tag{3.37}$$

Choose $C \ll C_1$ and $C \ll C_2$, then C_Σ can be approximated as

$$\frac{1}{C_\Sigma} \approx \frac{1}{C_q} + \frac{1}{C_0 + C} = \frac{C_q(C_0 + C)}{C_q + C_0 + C} \tag{3.38}$$

So the frequency is

$$f_0 = \frac{1}{2\pi\sqrt{L_q \frac{C_q(C_0 + C)}{C_q + C_0 + C}}} \tag{3.39}$$

(2) f_0 is always between f_s and f_p. Adjustment of capacitance C can make f_0 vary in a certain range. If C is set very large (suppose $C \to \infty$) from eq. (3.39), f_0 tends to approach the minimum boundary, which is the series resonant frequency f_s of crystal

$$f_{0\,min} \approx \frac{1}{2\pi\sqrt{L_q C_q}} = f_s \tag{3.40}$$

On the contrary, if C is set to be very small (suppose $C \approx 0$), then f_0 tends to approach the maximum boundary, which is the parallel resonant frequency f_p of crystal:

$$f_{0\,max} \approx \frac{1}{2\pi\sqrt{L_q \frac{C_q C_0}{C_q + C_0}}} = f_p \tag{3.41}$$

No matter how we adjust C, f_0 always falls into the range between f_s and f_p, validating that the crystal is inductive. Because of the crystal property in this region, Pierce oscillator is known to be highly stable in frequency.

(3) Fine-tuning of frequency. From the above analysis, we understand that the adjustment of C can change the frequency of oscillation. But why is fine-tuning necessary?

First of all, oscillation frequency of circuit sometimes is not exactly equal to the given nominal frequency of quartz crystal. There is tiny difference which needs to be corrected by the load capacitor C as shown in Fig. 3.21(b). The nominal frequency of crystal being 1 MHz, appropriate regulation of C from 24.5 pF to about 44.5 pF is needed so that it can make the actual oscillation frequency be 1 MHz.

Second, although quartz crystal is physically and chemically stable, capacitance or inductance still will drift with the environment (such as temperature) inevitably. Fine-tuning allows the circuit to maintain its frequency by manual or automatic adjustment despite the condition variation.

Example 3.2: The AC equivalent circuit of crystal oscillator in a digital frequency meter is shown in Fig. 3.22 with the working frequency of 5 MHz. Try to analyze the mechanism of the circuit and the function of crystal.

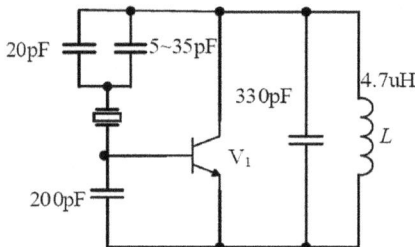

Fig. 3.22: Figure of Example 3.2.

Solution: There is a parallel LC loop (4.7 μH and 330 pF) between c–e electrodes of transistor V_1 and the resonant frequency:

$$f_0 = \frac{1}{2\pi\sqrt{4.7\times10^{-6}\times330\times10^{-12}}} \approx 4.0(\text{MHz})$$

At the working frequency of 5 MHz for the crystal, the LC loop is equivalent to a capacitor. The circuit can be seen as a Pierce oscillator, and the crystal is equivalent to an inductor. The variable capacitor 5–35 pF is for fine-tuning to make the oscillator operate at the nominal frequency of 5 MHz.

3.6.2.1.2 Miller oscillator

Miller oscillator is another quartz crystal oscillator working in parallel mode (Fig. 3.23). Quartz crystal is placed between the gate and the source electrodes of junction field-effect transistor (JFET) as an inductive component. LC circuit in parallel connected between the source and the drain is also inductive at the oscillation frequency. Interelectrode capacitance C_{gd} completes this inductance feedback type LC oscillator by being capacitive. C_{gd} is also known as Miller capacitance, so the circuit is named as Miller oscillator.

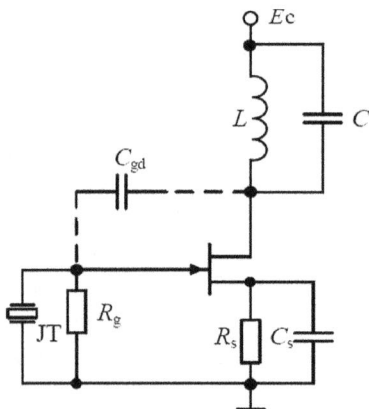

Fig. 3.23: Miller oscillator.

A field-effect transistor (FET) is chosen over a bipolar transistor in Miller oscillator. This is because the emitter junction resistance of forward-biased transistor is too small, which is bad for high Q-factor and may reduce the frequency stability. FETs are high in input impedance and therefore more preferred in this scenario.

3.6.2.1.3 Overtone crystal oscillator

Overtones of crystal oscillator are different from normal electrical harmonics. Electrical harmonics usually include the fundamental frequency and all integer higher harmonics (or multiples). These higher harmonics can coexist with the fundamental simultaneously. But the overtone of crystal oscillator which originated from its mechanical vibration only includes the fundamental and its odd multiples, with no coexistence of modes allowed.

The motivation of overtone oscillator application lies in higher frequency application. The nominal frequency of quartz crystal is determined by its geometry. As frequency goes higher the crystal wafer is required to be ever thinner, which is challenging for the mechanic tolerance. So the crystal oscillator works usually no more than 30 MHz due to the physical limit.

In order to employ crystals for higher working frequencies, we make the circuit oscillate at its overtones (usually at the third to the seventh harmonics), which is a special application called overtone crystal oscillator. In this way, we can use the crystal of fundamental frequency of tens of MHz to generate a stable oscillation of hundreds of MHz.

The schematic circuit of overtone crystal oscillator in parallel mode is shown in Fig. 3.24(a). What is different from Pierce oscillator is the L_1C_1 resonant circuit instead of a sole capacitor C_1 between c–e electrodes. According to the composition principle of feedback-type oscillators, the L_1C_1 resonant circuit should present capacitive to ensure positive feedback. Figure 3.24(b) shows the reactance characteristics of L_1C_1 loop. Integers on the horizontal axis refer to the order of harmonics.

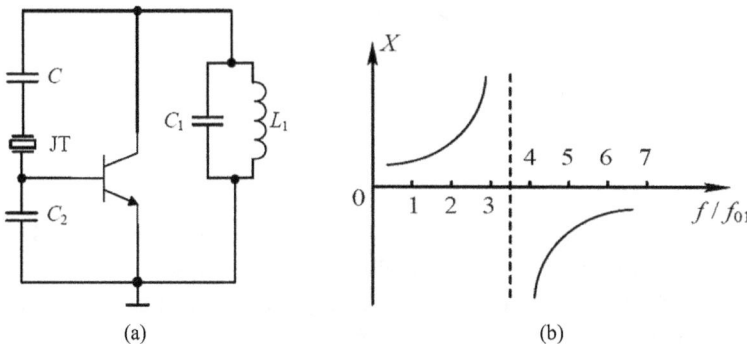

Fig. 3.24: Parallel mode overtone crystal oscillator: (a) schematic circuit and (b) reactance characteristics of $L1C1$ circuit.

From the graph we can get that the circuit cannot oscillate at the 3rd harmonic since L_1C_1 loop is inductive at that point. But at the 5th harmonic, L_1C_1 loop presents capacitive and meets the phase condition. If amplitude condition is satisfied at the same time, the circuit can oscillate at the 5th harmonic.

For 7th or higher orders overtone, although L_1C_1 loop presents capacitive, it is equivalent to a very large capacitance, making the capacitance divider ratio too small to meet the amplitude buildup conditions. Therefore, higher overtones usually can't be used for oscillation.

3.6.2.2 Crystal oscillators in series mode

For crystal oscillators work in series mode, the crystal works in the vicinity of its series resonant frequency f_s, with the equivalent impedance almost zero. Positive feedback loop is not "switched on" to produce oscillation unless frequency falls into the expected range ($\sim f_s$).

Block diagram of such a crystal oscillator is shown in Fig. 3.25. It is a two-stage common emitter amplifier, so the final output voltage is in inverse phase with the input voltage. The output is fed back to the input by the quartz resonator and load capacitor C_L, and the feedback is positive. Because of the frequency selectivity of the quartz resonator, only when the oscillating frequency is equal to the series resonant frequency determined by the quartz resonator and load capacitor C_L, the series impedance is the smallest and the positive feedback is the strongest. Oscillation occurs at this frequency.

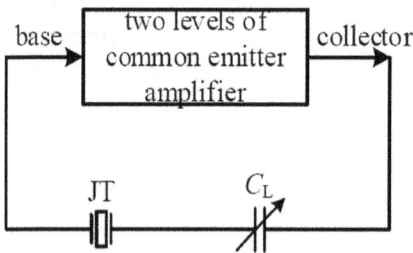

Fig. 3.25: Block diagram of crystal oscillator in series mode.

An actual circuit of crystal oscillator in series mode is given in Fig. 3.26. Quartz resonator is connected in series with the positive feedback circuit which is between two common emitter amplifiers. The collector load of the V_1 amplifier is a LC resonant circuit which is used to select frequency, and the better sinusoidal wave is obtained from the output end of V_2 amplifier. Adjusting the value of C_L can

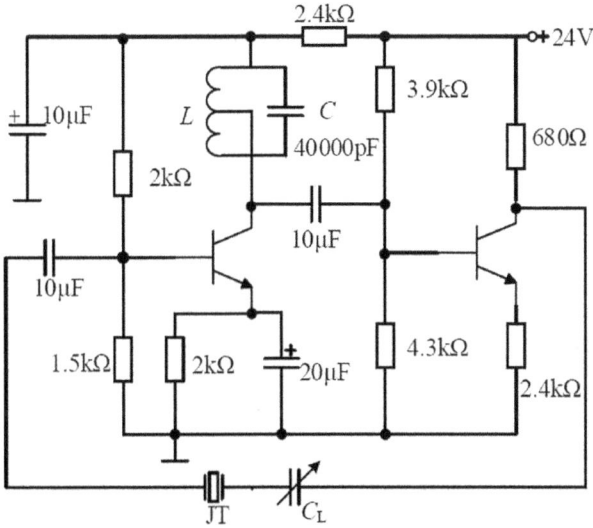

Fig. 3.26: Example circuit of crystal oscillator in series mode.

make the oscillation frequency overlap with the crystal nominal frequency. Different crystal choices can make the working frequency vary from tens of kHz to hundreds of kHz.

Another type of crystal oscillator in series mode resembles the capacitance feedback-type oscillators (Fig. 3.27). The only difference between the two is that the tap of capacitor divider first goes through a quartz resonator then back to the emitter electrode, to form a positive feedback pathway with frequency selectivity. C_1, C_2, C_3 and L constitute the resonant loop with the oscillation frequency tuned at series resonant frequency f_s of crystal, in which C_1 and C_2 are in parallel then in series with C_3.

When oscillation frequency equals f_s, the crystal presents purely resistive and the impedance is minimized. Under this condition, strongest positive feedback is achieved with zero phase shift. The frequency stability of oscillator depends on the quartz resonator. At other frequencies, the oscillation phase condition cannot be met.

3.7 Negative resistance oscillators

By energy conservation law, oscillations can sustain as long as the loss of oscillation can be compensated by the power supply. Negative resistance, discussed in this section, is also based on the energy conservation law. These are other examples of converting DC energy to AC energy, hence generating self-oscillation.

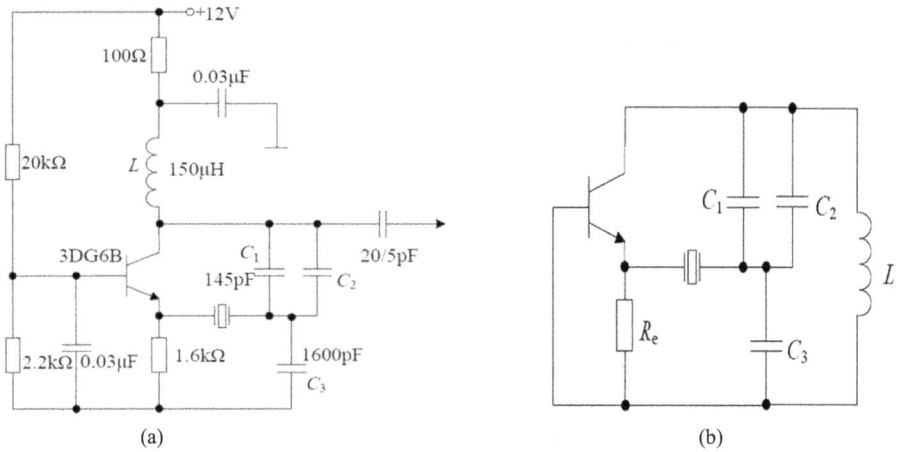

(a)

(b)

Fig. 3.27: Example circuit of crystal oscillator in series mode: (a) actual circuit and (b) AC equivalent circuit.

3.7.1 Basic characteristics of negative resistance device

Generally, both linear resistors and nonlinear resistors belong to positive resistors. Positive resistance is characterized as positively related current and voltage. To be more specific, the current going through the resistor and the power consumption both increase with increasing voltage across the resistor. Or inversely, voltage across the resistor increases with increasing current. This relationship can be expressed as

$$P = \Delta I \cdot \Delta U \tag{3.42}$$

where $\Delta U = R \cdot \Delta I$.

By contrast, negative resistance is described by negatively related current and voltage. Voltage across the resistor decreases as current going through increases.

$$\Delta U = -R \cdot \Delta I \tag{3.43}$$

Positive power indicates the consumption of energy, while negative power indicates the generation of energy. That is to say, under certain conditions negative resistance devices do not consume AC energy, but provide AC energy to the rest of circuit. Of course, AC energy does not come from nowhere which breaks the energy conservation law. The extra energy is obtained from DC power by energy conversion by negative resistance devices.

Figure 3.28 lists two types of negative resistance characteristics. Both i–u curves are negatively sloped in AB section. Depending on the curve shape Fig. 3.28(a) is called N-type negative resistance device and Fig. 3.28(b) is called S-type. N-type or voltage-controlled device is single valued function in current. Tunnel diode is an

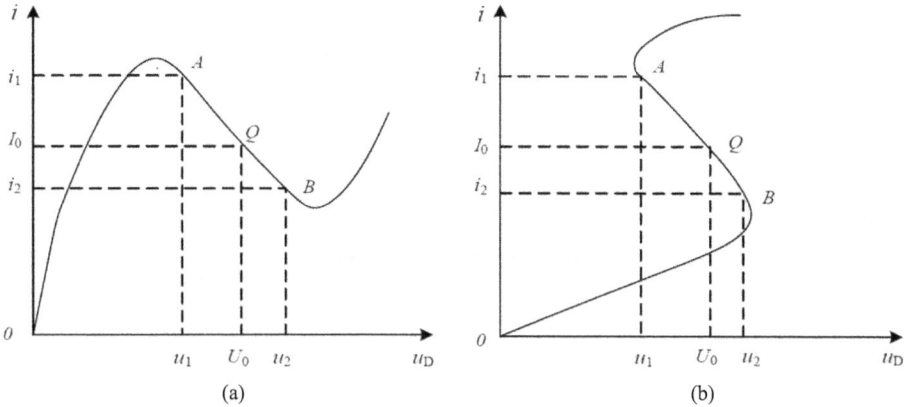

Fig. 3.28: Two typical characteristics of negative resistance devices: (a) N type and (b) S type.

example of N-type negative resistance device. S-type or current-controlled device is single valued in voltage. Unijunction transistor, silicon-controlled rectifier and so on fall into this category.

At present, a variety of novel negative resistance devices have constantly been invented and developed and they serve widely in practical circuits. For example, in the design of microwave solid state oscillator circuits, negative resistance oscillator can be used to compose avalanche diode and Gunn diode. The advantages of small size, lightness, low power consumption and high mechanic strength make negative resistance oscillators very valuable in electronic engineering. Moreover, they re-place the old fashioned microwave oscillators.

In this book, we will only focus on the illustration of principle of negative resis-tance oscillators by the external characteristic of negative resistance device. Now, we would discuss this question by an example of tunnel diode.

Operating status of a negative resistance device can be presented by DC parame-ters and differential parameters. In Fig. 3.28(a), if we choose point Q as the quiescent point of the device, then its DC resistance R and differential (or AC, or dynamic) resis-tance r can be shown as follow

$$R_0 = \frac{U_0}{I_0} \qquad (3.44)$$

$$r = \frac{\Delta u}{\Delta i} = -\frac{|U_2 - U_1|}{I_2 - I_1} \qquad (3.45)$$

It can be seen that its DC resistance is positive while AC resistance is negative. It dem-onstrates that the DC power is consumed while the AC power is generated. In other words, negative resistance device plays a role in converting energy from DC to AC.

However, DC offset is supplied to the DC power consumption of negative resistance device. Part of this power is converted to AC power, that is, the output power of negative resistance device to the output circuit, and the other part is the power consumption of device. The device with the negative characteristic cannot automatically produce AC power. So, we would constitute oscillating circuit by employing negative resistance device and enable it to obtain energy from DC power. Then, under the function of dynamic resistance, we can transform DC energy into AC power. This is the principle of the negative resistance oscillator.

3.7.2 Negative resistance oscillator circuits

Negative resistance oscillators are composed of negative resistance devices and frequency selection network. In order to ensure oscillation, current-controlled negative resistance device must be connected to series resonant circuit and voltage-controlled negative resistance device must be connected to parallel resonant circuit. The earliest negative resistance sinusoidal oscillator uses a tunnel diode as the negative resistance device. The principle of voltage-controlled negative resistance oscillator which is composed of tunnel diode will be introduced subsequently.

The circuit symbol and the AC equivalent circuit of a tunnel diode are shown in Fig. 3.29, where r_d is the negative dynamic resistance and C_d is interelectrode capacitance. Generally speaking, lead inductance and loss resistance are so small that they can be ignored in our circuit model.

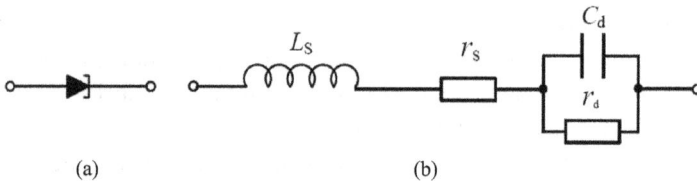

(a)   (b)

Fig. 3.29: (a) Tunnel diode symbol, (b) equivalent circuit.

The schematic circuit of tunnel diode oscillator is shown in Fig. 3.30(a). E_D and R_T provide stable bias circuit; L and C constitute the parallel resonant circuit; R_0 is the resonant resistance and R_L is the load resistance. If R_T is small enough and can be ignored, we can redraw the equivalent circuit as shown in Fig. 3.30(b). The effect of tunnel diode is replaced by parallel connected r_d and C_d, with the lead inductance and loss resistance ignored.

Suppose U is the effective voltage of oscillating loop circuit. Let $R_L' = R_L // R_0$, then the total power consumption on the load and the circuit is:

$$|P_L| = \frac{U^2}{R'_L} \tag{3.46}$$

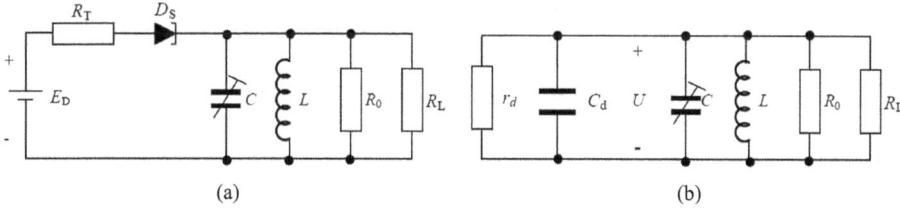

Fig. 3.30: A tunnel diode oscillator, (a) schematic circuit, (b) equivalent circuit.

The AC power supplied by r_d is

$$|P_d| = \frac{U^2}{r_d} \tag{3.47}$$

In order to ensure the circuit to start oscillating, the power generated by the negative resistance must be greater than the power consumed by the load and the circuit resistance. This requirement is expressed as

$$|P_d| > |P_L| \tag{3.48}$$

which leads to

$$R'_L > r_d \tag{3.49}$$

This is the condition for the circuit to start oscillation, and r_d is related to the intensity of oscillation. Since the characteristics of negative resistance is nonlinear, r_d increases with the increasing amplitude until it meets the equilibrium condition of

$$R'_L = r_d \tag{3.50}$$

Under the equilibrium state, positive resistance and negative resistance effects are 100% compensated in maintaining a stable oscillation amplitude.

Apparently, the oscillation frequency of this circuit is approximately equal to the resonant frequency of LC parallel circuit given by

$$f_0 \approx \frac{1}{2\pi\sqrt{L(C + C_d)}} \tag{3.51}$$

In general, bias voltage of the working point of tunnel diode is about 100 or 200 mV and E_D is required to be kept very low. As it is not convenient in circuit realization, we can alternatively use a larger E_D with voltage division (Fig. 3.31(a)). Additionally, a capacitor C_1 is parallel connected with the tunnel diode in order to reduce the influence of diode capacitance C_d instability (Fig. 3.31(a)). In Fig. 3.31(b), R is equivalent

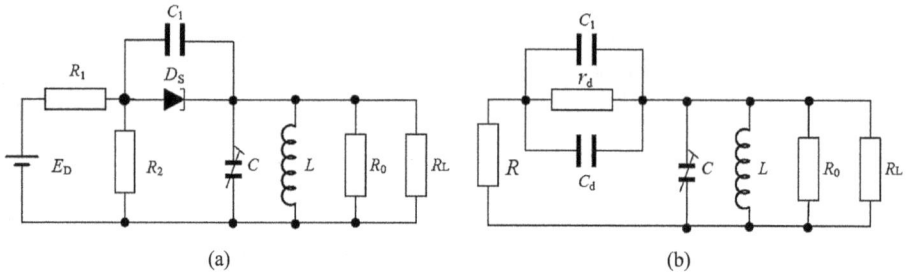

Fig. 3.31: Tunnel diode oscillator circuit: (a) actual circuit and (b) AC equivalent circuit.

resistance of R_1 and R_2. R is large and not negligible. When calculating power consumption, we should take both $R_L{}'$ and R into consideration. In addition, the highest frequency is restrained by capacitance of C_1.

Although tunnel diode oscillator circuit looks simple, the structure is vitally important when it is applied in the microwave field. Resonant cavity or strip line is generally used as the resonant circuit. The operational frequency can be as high as several thousand MHz with the advantages of small size and low power consumption. The disadvantage is mainly small output power. Now as the microwave technology advances further, new developed negative resistance oscillators can overcome this shortcoming and enables much more extensive applications.

3.8 Voltage-controlled oscillators

The reactance of certain electrical component can be controlled by an applied voltage. If the variable reactance component is connected with an oscillator, then the oscillation frequency becomes tunable by the applied voltage. This is a brief explanation of mechanism of voltage-controlled oscillators (VCO). One of the most frequently used components in a VCO is a varactor diode.

VCOs are widely used in frequency modulation, frequency synthesis, phase locked loop circuits, TV tuner and spectrum analyzer and so on.

3.8.1 Varactor diode

Varactor diode is a voltage-controlled reactance component. The junction capacitance of its PN junction can dramatically change along with the reversely applied voltage. The varactor symbol and the reverse-biased junction capacitance curve are shown in Fig. 3.32.

The capacitance C_j of a reverse-biased PN junction is a nonlinear function of applied junction voltage u known as

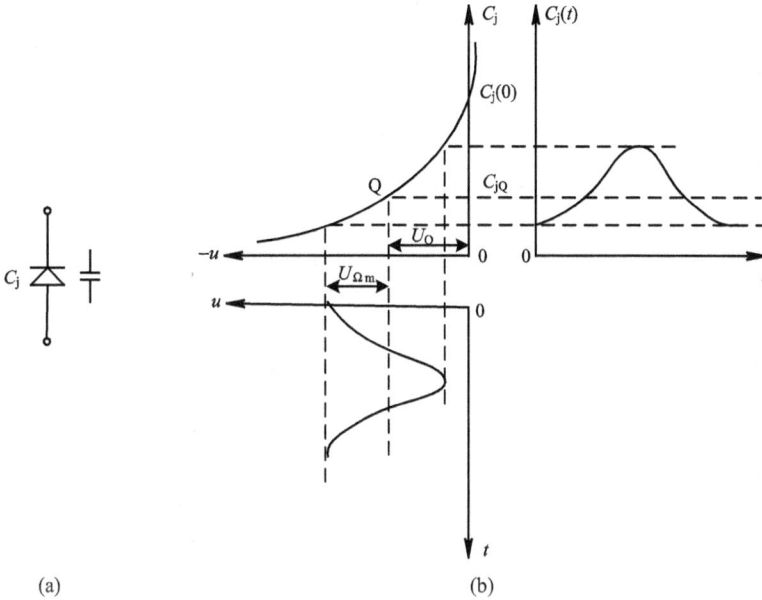

Fig. 3.32: Varactor, (a) symbol, (b) reverse-biased junction capacitance curve versus voltage.

$$C_j = \frac{C_j(0)}{\left(1 - \frac{u}{U_B}\right)^n} \tag{3.52}$$

where $C_j(0)$ is the diode capacitance at zero bias ($u = 0$), U_B is the contact potential of PN junction and n is the variable capacitance index depending on the semiconductor doping concentration and the structure of PN junction. For the ideal mutant PN junction, n is typically 1/2, and if the PN junction is approximately linear, n is 1/3.

The varactor must work in the reverse bias state, so it is necessary to add a negative DC bias voltage U_Q as the quiescent point (Q point). If the AC voltage is in the cosine form, then the voltage on the varactor can be written as

$$u = -(U_Q + u_\Omega) = -(U_Q + u_{\Omega m} \cos \Omega t) \tag{3.53}$$

By substituting that into eq. (3.52), we obtain

$$C_j = \frac{C_{jQ}}{\left(1 + \frac{u_\Omega}{U_B + U_Q}\right)^n} = \frac{C_{jQ}}{(1 + m \cos \Omega t)^n} \tag{3.54}$$

where the static junction capacitance at Q point is

$$C_{jQ} = \frac{C_j(0)}{\left(1 + \frac{U_Q}{U_B}\right)^n} \tag{3.55}$$

The modulation index is

$$m = \frac{U_{\Omega m}}{U_B + U_Q} < 1 \qquad (3.56)$$

3.8.2 Varactor diode voltage-controlled oscillators

When the varactor is connected into an LC oscillator, a VCO can be formed. We can transform various feedback type LC oscillators to VCOs.

To make the varactor work normally, the negative DC bias voltage and AC control voltage should be properly applied, and the high-frequency oscillation signal interfering the DC bias and control voltage should be prevented, and RF choke, bypass capacitors and blocking capacitors should exist in the VCO circuit design.

The analysis of a VCO requires readers to firstly find out the DC or low-frequency pathways and the high-frequency pathways. This analyzing skill is essential for future study of frequency modulation circuits as well. From the DC path, we can understand the static parameters and from high-frequency path; we can judge the type of oscillator and study the RF performance. Next we are going to study an example and practice.

Example 3.3: The varactor diode VCO with center frequency 100 MHz is shown in Fig. 3.33(a). Please draw the DC path and high-frequency oscillation circuit of transistor, as well as the varactor's DC bias circuit and low-frequency control circuit.

Solution: To draw the transistor's DC path as shown in Fig. 3.33(b), capacitors should be considered as open, and inductors should be short. The varactor diode should also be open, since it works in the reverse bias state.

To draw the DC bias circuit of the varactor diode as shown in Fig. 3.33(c), all capacitors related to the varactor diode should be open, and the inductors should be short.

Before drawing the high-frequency oscillation circuit and low-frequency control circuit, each capacitor and inductor should be carefully analyzed. Usually for the high-frequency oscillation circuit, small capacitors contribute to oscillation are working capacitors, and significantly large capacitors are either coupling capacitors or bypass capacitors. Small inductors contribute to oscillation are working inductors, and significantly large inductors serve as RF choke. Of course, the varactor is also a working capacitor.

By keeping the working capacitors and inductors in position, shorting all the coupling and bypass capacitors, opening the RF choke and grounding the DC power supply, we can get the high-frequency oscillation circuit as shown in Fig. 3.33(d). It can be seen that it is a capacitance feedback-type LC VCO.

To draw the low-frequency control loop, we first need to short the inductor and RF choke connected with the varactor diode (inductance or RF choke is small at low frequency) and ground the DC power supply. Except for the large coupling and bypass capacitors (considered short even at low frequency), the other small capacitors are considered as open. Then, we can get the low-frequency control loop as shown in Fig. 3.33(e).

Obviously, if the control voltage u_Ω is not given, the oscillation frequency of VCO will change with the change of adjustable resistance R_2.

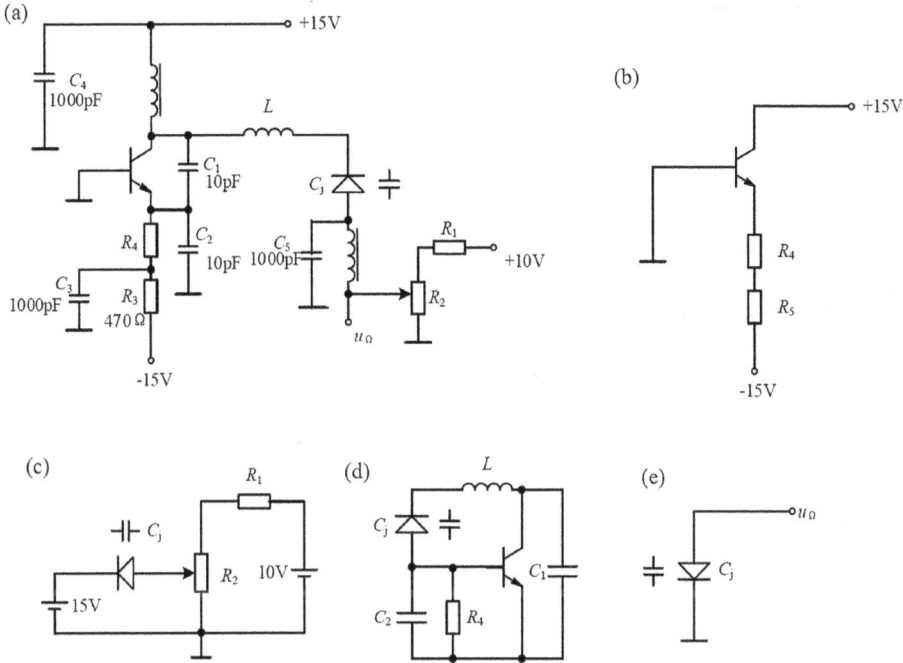

Fig. 3.33: Example 3.3.

3.8.3 RF CMOS voltage-controlled oscillators

In recent years, CMOS inductance–capacitance resonant fully integrated VOC (*LC-VCO*) has been applied in the RF unit circuit of wireless transceiver in the academic and industrial research. The most important features of VCO include low phase noise, low power consumption and wide tuning range.

In order to tune the *LC* oscillator frequency, the resonance parameters must be adjusted accordingly. Instead of changing inductance which is inconvenient, we choose to vary the capacitance by means of a varactor. MOS varactors are more commonly used over PN junctions, especially in low-voltage design. We thus construct the VCO circuit as shown in Fig. 3.34, where varactors M_{U1} and M_{U2} appear in parallel with the tanks (assume U_{cont} is provided by an ideal voltage source). Note that the gates of varactors are tied to the oscillator nodes and the source/drain/n-well terminals to U_{cont}. This avoids loading X and Y with the capacitance between n-well and substrate.

Since the gates of M_{U1} and M_{U2} reside at an average level equal to E_D, their gate-source voltage remains positive and their capacitance decreases as U_{cont} goes from zero to E_D as shown in Fig. 3.35. This behavior persists even in the presence of large voltage swings at X and Y and hence across M_{U1} and M_{U2}. The key point is that the

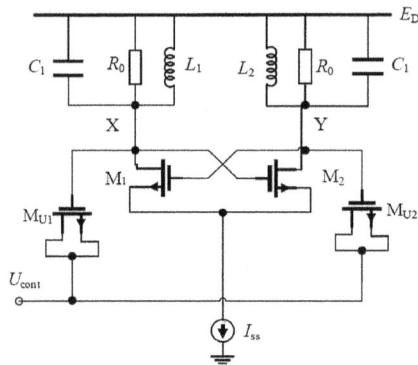

Fig. 3.34: Schematic of VCO with MOS varactors.

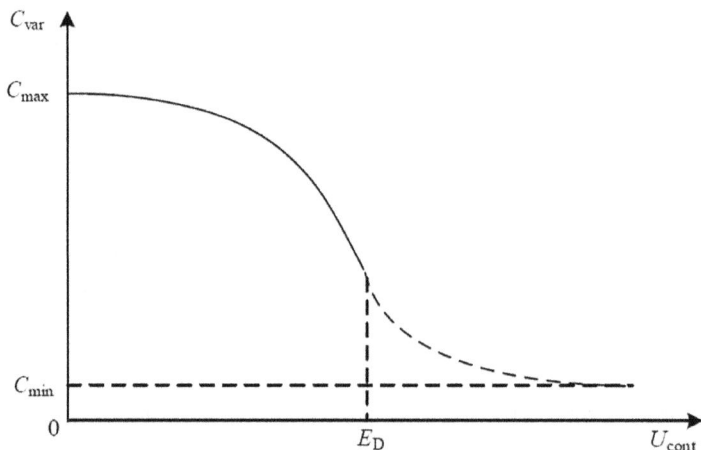

Fig. 3.35: Varactor capacitance versus controlled voltage.

average voltage across each varactor goes from E_D to zero as U_{cont} goes from zero to E_D, thus creating a monotonic decrease in their capacitance.

The oscillation frequency can thus be expressed as

$$\omega_{osc} = \frac{1}{\sqrt{L_1(C_1 + C_{var})}} \tag{3.57}$$

where C_{var} denotes the average value of each varactor's capacitance.

Why the capacitor C_1 should be included in the oscillator shown in Fig. 3.34? It appears that without C_1, the varactors can vary the frequency to a greater extent, and thereby C_1 can provide a wider tuning range. This is indeed true, and we rarely need to add a constant capacitance to the tank deliberately. In other words, C_1 simply models the inevitable capacitances appearing at X and Y: (1) C_{GS}, C_{GD} (with a Miller multiplication factor of 2) and C_{DB} of M_1 and M_2; (2) the parasitic capacitance of each inductor; (3) the input capacitance of the next stage (e.g., a buffer, divider or mixer). The last component becomes particularly significant on the transmit side due to the "propagation" of capacitance from the input of PA to the input of up-conversion mixers.

3.9 Summary

In this chapter we discussed oscillators from the theoretical principle and conditions, the most basic LC oscillator circuits (Colpitts, Hartley, Clapp and Seiler), then quartz crystal oscillators and finally negative resistance oscillators and VCOs. For all circuits introduced, the logical flow tries to maintain the consistency as follows: principle, circuit composition, oscillation condition analysis, oscillation parameters calculation, advantages and disadvantages, possible improvement discussion and finally practical examples.

We hope the readers could understand the mechanism and circuit design of various oscillators, master the analysis technique through the learning process and are capable of related calculations.

Problems

3.1 The capacitance feedback oscillator circuit is shown in Ex.Fig. 3.1, $C_1 = 100$ pF, $C_2 = 200$ pF and $L = 30$ μH, calculate the oscillating frequency and the minimum value of necessary amplifier gain K_{min} to maintain the oscillation.

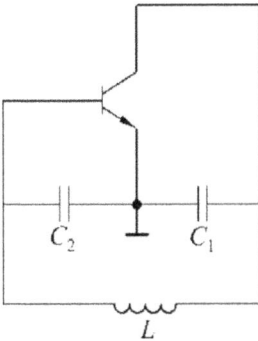

Ex.Fig. 3.1

3.2 By using the phase equilibrium condition, determine which of the oscillation circuits shown in Ex.Fig. 3.2 are possible to oscillate, specify the type of oscillation circuits and explain the oscillation conditions.

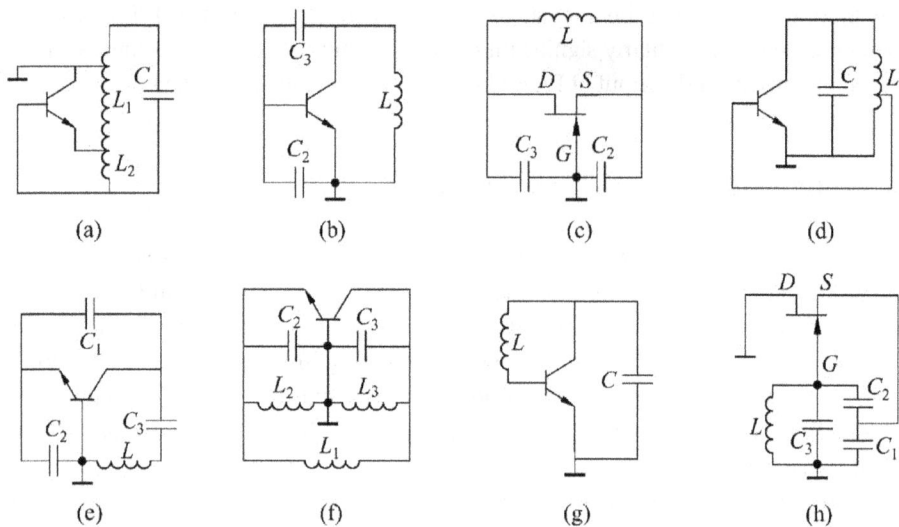

(a)   (b)   (c)   (d)

(e)   (f)   (g)   (h)

Ex.Fig. 3.2

3.3 The AC equivalent circuit of three-loop oscillator is shown in Ex.Fig. 3.3, assuming the following six conditions:
(1) $L_1C_1 > L_3C_3 > L_2C_2$
(2) $L_1C_1 < L_2C_2 < L_3C_3$
(3) $L_1C_1 = L_2C_2 = L_3C_3$
(4) $L_1C_1 = L_3C_3 > L_2C_2$
(5) $L_1C_1 < L_2C_2 = L_3C_3$
(6) $L_2C_2 > L_3C_3 > L_1C_1$

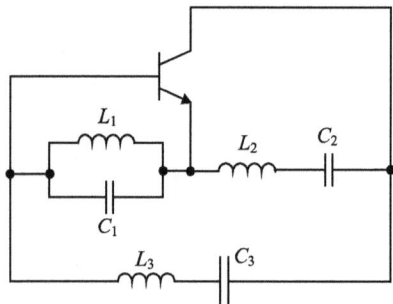

Ex.Fig. 3.3

Which are the valid conditions for oscillation? What is the equivalent type of oscillation circuit? What is the relationship between oscillation frequency and the natural frequency of each loop?

3.4 Two oscillators are shown in Ex.Fig. 3.4, with the resonant frequencies of two LC parallel resonant circuits being $f_1 = \frac{1}{2\pi\sqrt{L_1C_1}}$ and $f_2 = \frac{1}{2\pi\sqrt{L_2C_2}}$. Try to explain the relationship between oscillation frequency f_0 and f_1, f_2, and specify the types of oscillator.

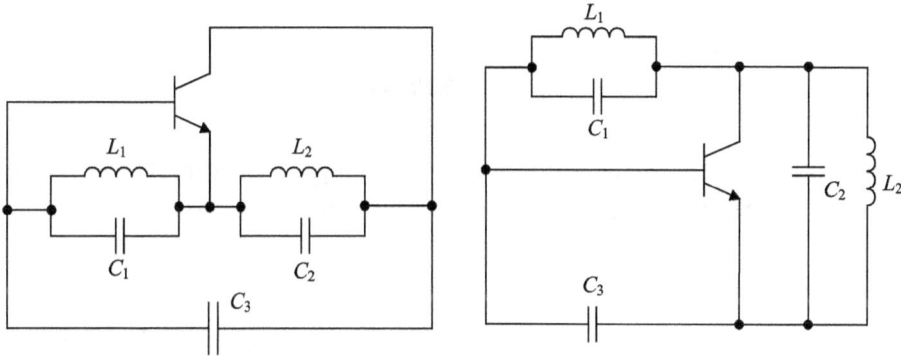

Ex.Fig. 3.4

3.5 The oscillation circuit is shown in Ex.Fig. 3.5.
   (1) Draw the AC equivalent circuit.
   (2) Calculate the oscillating frequency f_0 and feedback coefficient F.

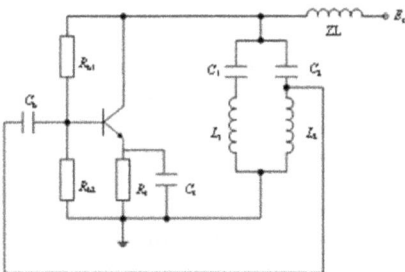

Ex.Fig. 3.5

3.6 The oscillation circuit is shown in Ex.Fig. 3.6.
   (1) Draw the high-frequency AC equivalent circuit, and explain the type of oscillator.
   (2) Calculate the oscillating frequency f_0.

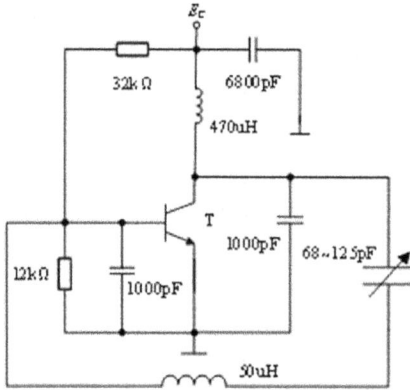

Ex.Fig. 3.6

3.7 The crystal oscillator of 25 MHz is shown in Ex.Fig. 3.7. Draw the AC equivalent circuit, explain the role of crystal and calculate the feedback coefficient F.

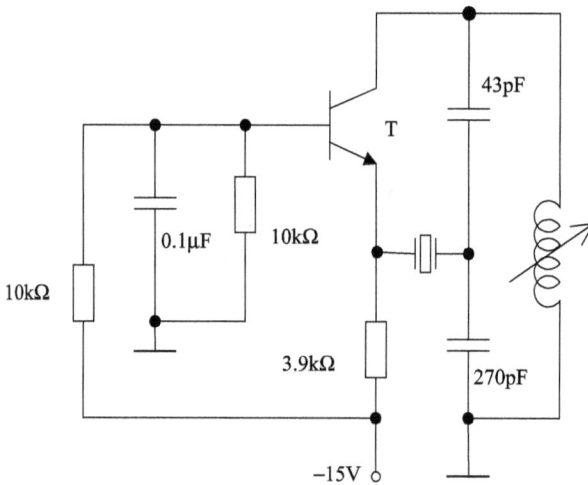

Ex.Fig. 3.7

3.8 Draw a practical crystal oscillation circuit to meet the following requirements:
   (1) Use a transistor.
   (2) The crystal is used as an inductor.
   (3) The LC parallel circuit is between the collector and the emitter of transistor.
   (4) The emitter of transistor is AC grounded.

3.9 The crystal oscillation circuit is shown in Ex.Fig. 3.8. The known crystal and C_1 constitute a parallel resonant circuit and its resonant resistance $R_0 = 80$ kΩ, $R_f / R_1 = 2$.
   (1) Analyze the function of crystal.

(2) To meet the startup conditions, what is the maximum possible value of R_2? (Assuming the integrated operational amplifier is ideal.)

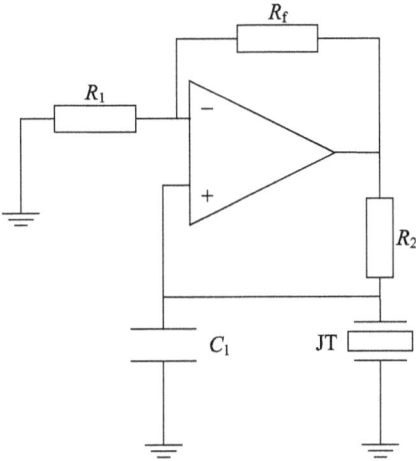

Ex.Fig. 3.8

3.10 The third overtone crystal oscillator of output oscillation frequency 5 MHz is shown in Ex.Fig. 3.9, try to draw the high-frequency AC equivalent circuit, and explain the role of LC circuit(L is 4.7 μH and C is 330 pF).

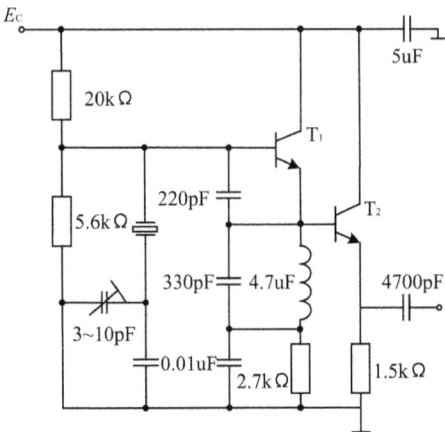

Ex.Fig. 3.9

Chapter 4
Amplitude modulation, demodulation and frequency mixing

4.1 Introduction

Modulation circuits and demodulation circuits are important parts of communication system. As introduced in Chapter 1, in order to transmit signals from a transmitter to a receiver, modulation and demodulation must be performed.

Transmission of multiple signals at the same time is required further as a more efficient communication style. Practically, a typical wireless transmission system includes two types of signals: a high-frequency carrier signal and a low-frequency information signal, which is loaded to the carrier signal by a transmitter and removed by a receiver.

In communication electronic circuits, modulation means tuning a higher frequency continuous carrier signal with a modulation wave-carrying information, such as an audio signal for a sound, or a video signal for an image. So the carrier wave can carry information. When the modulated carrier reaches its destination, the information signal will be extracted by a demodulation process.

So we can summarize why we need modulation:
(1) Make the actual wireless transmission of signals be possible.
(2) Use the carrier frequency to realize the power-efficient transmission.
(3) Embed the low-frequency information into the high-frequency carrier.

According to the adopted carrier waveform, modulation can be divided into continuous-wave (sine wave) modulation and pulse modulation. This book deals only with sine wave modulation. The sine wave modulation is based on the high-frequency sine wave as carrier, and the low-frequency modulation (FM) signal is used to control the amplitude, frequency or phase of sine wave. So, there are three basic and classic modulation methods: amplitude modulation (AM) (introduced in this chapter), FM and phase modulation (PM) (introduced in Chapter 5).

AM is a modulation method for communication electronic circuits, which is usually used to transmit information through radio waves. Today, it still exists in many forms of communication electronic devices. For instance, it is still used for portable radios, mobile phones and computers. "AM" often refers to medium-wave AM radio broadcasting.

In AM, the carrier wave's amplitude (signal strength) varies with the low-frequency waveform being transmitted. For example, in AM radio, the amplitude of a sinusoidal continuous wave is modulated by an audio signal before transmission. The amplitude of carrier wave, or the envelope of waveform, is modified by the audio

https://doi.org/10.1515/9783110593822-004

waveform. On the contrary, FM allows the carrier signal's frequency being varied, and PM allows the carrier signal's phase being varied.

Demodulation is the contrary process of the modulation, and the low FM signals can be removed from the high-frequency carrier. Correspondingly, there are also three demodulation methods: amplitude demodulation (introduced in this chapter), frequency demodulation and phase demodulation (introduced in Chapter 5).

The essence of modulation and demodulation is frequency transformation, which is characterized by new frequency components that differ from the input signal produced in the output signal. The aim of frequency transformation is to transform the spectrum of input signal, in order to get the desired spectrum of output signal.

Frequency transformation is divided into two categories: linear shift of spectrum and nonlinear transformation of spectrum. AM, demodulation and frequency mixing belong to the linear shift of spectrum. FM and demodulation, PM and demodulation all belong to the nonlinear transformation of spectrum.

In order to realize the frequency conversion, the nonlinear electronic device such as a diode, a transistor or a field-effect transistor (FET) must be used. In recent years, integrated circuits have been widely used in analog communication. Modulators, demodulators and mixers can be realized by integrated analog multipliers.

In this chapter, we first introduce the basic characteristics of AM and demodulation in time domain and frequency domain, respectively, and then introduce the related circuit composition and operational principle. Frequency mixing belongs to the linear shift of spectrum as well as AM and demodulation, so it is also introduced in this chapter.

4.2 Analysis of amplitude-modulated signal

AM is to use a low-frequency modulating signal to control the amplitude of a high-frequency carrier wave, so that the amplitude of carrier wave varies proportionally with the modulating signal. The high-frequency carrier whose amplitude is modulated is called amplitude-modulated wave.

According to the different output spectrum structure, the amplitude-modulated wave can be divided into three types: standard AM wave (AM), double-sideband/suppressed-carrier AM wave (DSB/SC-AM) and single-sideband/suppressed-carrier AM wave (SSB/SC-AM).

4.2.1 Standard AM wave

Theoretically, standard AM is the simplest method of AM. In this method, the envelope of standard AM wave changes linearly with the modulation signal, so the shape of base-band information-carrying modulation signal (e.g., voice signal) is reproduced with the amplitude of carrier signal. A complex signal consists of multiple singles. First, the analytical expression of standard AM signal is derived.

4.2.1.1 Mathematical expressions of standard AM wave

Assume that the low FM signal is a monophonic audio signal:

$$u_\Omega(t) = U_{\Omega m} \cos \Omega t = U_{\Omega m} \cos 2\pi F t \qquad (4.1)$$

The high-frequency carrier signal is

$$u_c(t) = U_{cm} \cos \omega_c t = U_{cm} \cos 2\pi f_c t \qquad (4.2)$$

where $U_{\Omega m}$ is the maximal amplitude of the modulation signal, F is the modulation frequency, U_{cm} is the maximal amplitude of a carrier and f_c is carrier frequency. To simplify the analysis, let the initial phase angles of above two signals be 0.

Because the amplitude of standard AM wave is proportional to the modulation signal and the carrier frequency is constant after modulation, the expression of standard AM wave is

$$u_{AM}(t) = (U_{cm} + k_a U_{\Omega m} \cos \Omega t) \cos \omega_c t$$

$$= U_{cm}\left(1 + k_a \frac{U_{\Omega m}}{U_{cm}} \cos \Omega t\right) \cos \omega_c t \qquad (4.3)$$

$$= U_{cm}(1 + m_a \cos \Omega t) \cos \omega_c t$$

where the AM index is defined as $m_a = k_a U_{\Omega m}/U_{cm}$, which indicates the extent to which the carrier amplitude is controlled by the modulation signal, and k_a is a proportional constant determined by the AM circuit.

4.2.1.2 Waveform of standard AM wave

Based on expression (4.3), we can obtain the waveform of standard AM wave, as shown in Fig. 4.1. It can be seen that the standard AM wave is also a high-frequency oscillation, and its amplitude variation (i.e., envelope variation) is completely consistent with the modulation signal. Therefore, the standard AM wave carries the information of the original modulation signal.

From Fig. 4.1, we can find the maximum value of envelope U_{max} and the minimum value of envelope U_{min}:

$$U_{max} = U_{cm}(1 + m_a) \qquad (4.4)$$

$$U_{min} = U_{cm}(1 - m_a) \qquad (4.5)$$

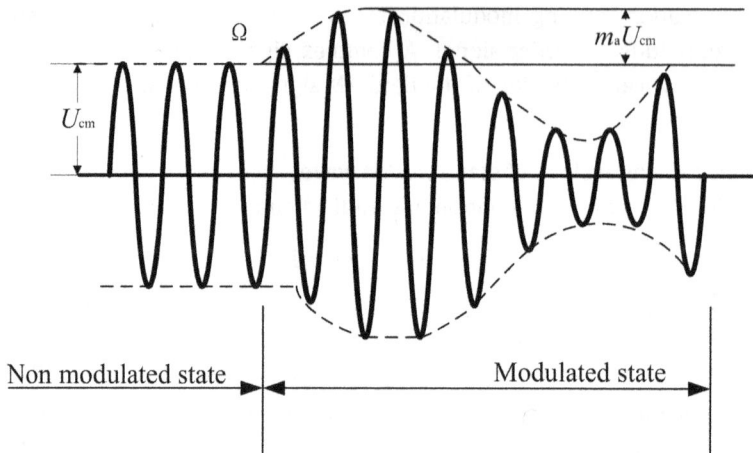

Fig. 4.1: Waveform of standard AM wave.

So, we can obtain the AM index m_a:

$$m_a = \frac{U_{max} - U_{min}}{U_{max} + U_{min}} \tag{4.6}$$

The AM index m_a is an important modulation parameter that shows the ratio of base-band maximum amplitude and carrier maximum amplitude (Fig. 4.1). When the carrier is not modulated, $m_a = 0$; when the m_a value is bigger, the more modulation signal is embedded in the envelope; when $m_a = 1$, the maximum m_a value is reached, which is known as 100% standard AM wave. For effective power transmission and high SNR (signal-to-noise ratio), it is expected to have the amplitude of modulation signal as high as possible relative to the carrier's amplitude.

Figure 4.2 shows the waveform of standard AM wave with different m_a values. If the maximal amplitude of carrier is greater than the amplitude of the modulation signal, that is, $m_a = 0.5$, then the embedded envelope is a faithful representation of the information (in this case, it is a clean sinusoidal shape). If the maximal amplitude of carrier is equal to the amplitude of modulation signal, that is, $m_a = 1$, then the embedded envelope is still a faithful copy of the information.

When $0 < m_a \leq 1$, the standard AM signal has two symmetrical envelopes, one positive and the other negative, with the same information. As long as the two envelopes remain separate without overlapping (positive envelope remains positive and negative envelope remains negative), the information can be recovered from either of the two envelopes.

However, when the modulation index $m_a > 1$, the envelope is not a loyal copy of the information. Once the two envelopes overlap (Fig. 4.2), the message is distorted because the portion of positive envelope crosses and becomes part of the negative

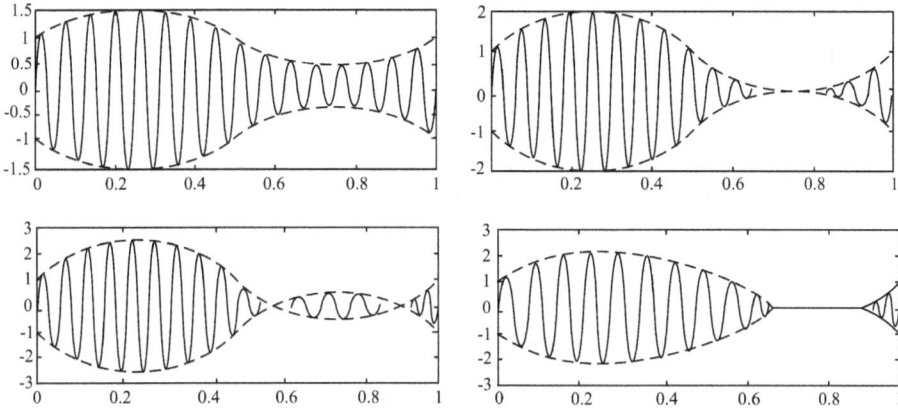

Fig. 4.2: Time domain plots of different AM index $m_a = 0.5, 1, 1.5$.

envelope, which results in a signal clip. This is called "over-modulation": the positive and the negative envelopes do not look like the original information (it looks like a clipped sine wave). Note that in the case of overmodulation, eq. (4.3) is invalid in the crop area.

4.2.1.3 Frequency spectrum of standard AM wave

By inspecting eq. (4.3), we recognize that AM is equivalent to multiplication in time domain, and we can further obtain

$$u_{AM}(t) = U_{cm}(1 + m_a \cos \Omega t) \cos \omega_c t$$

$$= U_{cm} \cos \omega_c t + \frac{1}{2} m_a U_{cm} \cos(\omega_c + \Omega)t + \frac{1}{2} m_a U_{cm} \cos(\omega_c - \Omega)t$$

(4.7)

Therefore, the frequency content of standard AM wave contains three high-frequency components: the carrier frequency ω_c, the upper side frequency $(\omega_c + \Omega)$ and the lower side frequency $(\omega_c - \Omega)$. It is important to notice that the amplitude of side frequency components is multiplied by $m_a/2$, that is, in the best case of $m_a = 1$, the amplitude of side frequency components is half of the amplitude of carrier.

By drawing these three frequency components, the spectrum diagram of standard AM wave shown in Fig. 4.3 can be obtained.

The above analysis shows that the process of standard AM is to move the low FM signal to the two sides of high-frequency carrier component in the spectrum. That is, in frequency domain, it is a linear spectrum shifting process. Apparently in

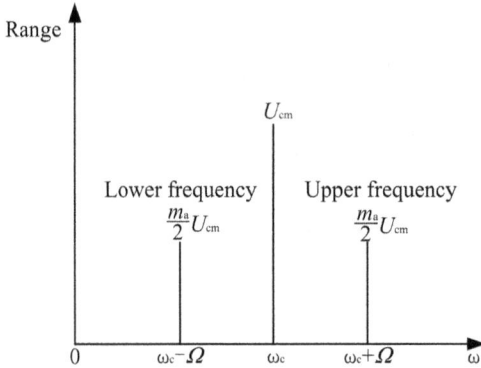

Fig. 4.3: Frequency spectrum of standard AM wave.

the standard AM wave, carrier does not contain any useful information, and the information to be transmitted is present only in the two sidebands.

We also observe that for a monophonic audio modulation signal whose frequency is Ω, the standard AM wave occupies the bandwidth BW = 2 Ω that is centered around the carrier frequency.

In fact, the modulation signal is not a sinusoidal signal of a single frequency, but a complex waveform containing several frequency components. In multi-FM, if modulated by a number of signals of different frequencies$\Omega_1, \Omega_2, \ldots, \Omega_{max}$, the standard AM wave equation is

$$u_{AM}(t) = U_{cm}(1 + m_{a1}\cos\Omega_1 t + m_{a2}\cos\Omega_2 t + \cdots + m_{an}\cos\Omega_{max}t)\cos\omega_c t$$

$$= U_{cm}\cos\omega_c t + \frac{m_{a1}}{2}U_{cm}\cos(\omega_c + \Omega_1)t + \frac{m_{a1}}{2}U_{cm}\cos(\omega_c - \Omega_1)t$$

$$+ \frac{m_{a2}}{2}U_{cm}\cos(\omega_c + \Omega_2)t + \frac{m_{a2}}{2}U_{cm}\cos(\omega_c - \Omega_2)t + \cdots \qquad (4.8)$$

$$+ \frac{m_{an}}{2}U_{cm}\cos(\omega_c + \Omega_{max})t + \frac{m_{an}}{2}U_{cm}\cos(\omega_c - \Omega_{max})t$$

Accordingly, it contains a carrier frequency component and a series of upper and lower frequency components. The total bandwidth is twice the maximum modulation frequency, that is, BW = 2 F_{max}.

4.2.1.4 Average power of standard AM wave
If the standard AM wave voltage is applied to the load resistance R_L, the power absorbed by the load resistance is the sum of power when each sinusoidal component

acts alone. Based on eq. (4.7), we can write out the power obtained on R_L, which consists of three parts:

Carrier power

$$P_c = \frac{U_{cm}^2}{2R_L} \qquad (4.9)$$

Upper side frequency component power

$$P_1 = \frac{1}{2}\left(\frac{m_a}{2}U_{cm}\right)^2 \frac{1}{R_L} = \frac{1}{8}\frac{m_a^2 U_{cm}^2}{R_L} = \frac{1}{4}m_a^2 P_c \qquad (4.10)$$

Lower side frequency component power

$$P_2 = \frac{1}{2}\left(\frac{m_a}{2}U_{cm}\right)^2 \frac{1}{R_L} = \frac{1}{8}\frac{m_a^2 U_{cm}^2}{R_L} = \frac{1}{4}m_a^2 P_c \qquad (4.11)$$

Therefore, the average power given by the standard AM wave in one period of the modulation signal is

$$P = P_c + P_1 + P_2 = P_c\left(1 + \frac{m_a^2}{2}\right) \qquad (4.12)$$

It can be seen that the side frequency power increases with the increase of m_a. When $m_a = 1$, the side frequency power is maximum, and the maximum total average power required for $m_a = 1$ is $P = 1.5P_c$, while each of the sidebands is only $1/4P_c$. Our conclusion is that even for 100% standard AM wave, that is, $m_a = 1$, each sideband contains only 1/6 of the total power (each sideband contains its own copy of useful information), while 2/3 of the total power is taken by the carrier (which does not contain any information).

In other words, with the standard AM, the power transmitted by the transmitter is mostly taken up by carriers without information. Obviously, this is very uneconomical. However, because the modulation equipment is simple, especially the demodulation is simpler and easy to receive, it is still widely used in some fields.

Although the above analysis focused on monophonic signals, we should remember that nonsinusoidal modulation signals are composed of several sine waves, not necessarily related to harmonics. The overall average power is the sum of average power of single tones:

$$P = P_c\left(1 + \frac{m_{a1}^2}{2} + \frac{m_{a2}^2}{2} + \cdots + \frac{m_{an}^2}{2}\right) \qquad (4.13)$$

where m_{ai} ($i = 1,2, \ldots ,n$) is the modulation index of single tone i.

4.2.2 Double-sideband AM wave

Because the carrier does not carry information, in order to save transmitting power, we can only transmit the upper and lower sideband signals carrying information, and the carrier with no information can be suppressed in transmission. This modulation method is called DSB/SC-AM.

DSB/SC-AM signal can be obtained by multiplying modulation signal and carrier signal directly, which can be written as

$$u_{DSB}(t) = Au_\Omega u_c = AU_{\Omega m}\cos\Omega t \cdot U_{cm}\cos\omega_c t$$

$$= \frac{1}{2}AU_{\Omega m}U_{cm}[\cos(\omega_c + \Omega)t + \cos(\omega_c - \Omega)t]$$

$$(4.14)$$

where A is the coefficient determined by the AM circuit.

Based on eq. (4.14), $AU_{\Omega m}U_{cm}\cos\Omega t$ is the amplitude of a DSB signal, which is proportional to the amplitude of modulation signal. The amplitude of high-frequency carrier signal varies according to the regulation of modulation signal, not on the basis of U_{cm}, but on the basis of zero, which can be positive and negative.

Therefore, the phase of corresponding high-frequency carrier oscillation has a sudden change of 180° when the modulation signal enters the negative half cycle from the positive half cycle (i.e., the zero-crossing point of AM envelope).

The waveform of DSB signal is shown in Fig. 4.4. It can be seen that the envelope of DSB amplitude-modulated wave no longer reflects the changing law of modulation signal.

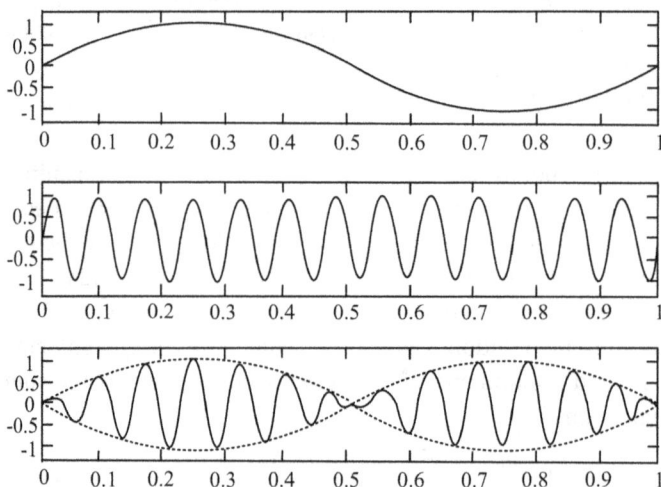

Fig. 4.4: Waveform of DSB amplitude-modulated signal.

The frequency spectrum of DSB/SC-AM signal is shown in Fig. 4.5. It can be seen that the signal is still concentrated near the carrier frequency, keeping the characteristic of linear spectrum shift. The bandwidth occupied is BW = 2F.

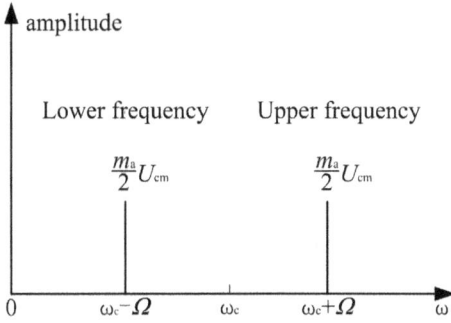

Fig. 4.5: Frequency spectrum of DSB/SC-AM signal.

4.2.3 Single-sideband AM wave

Further observing the frequency spectrum of DSB/SC-AM signal, we can find that both the upper sideband and the lower sideband reflect the spectral structure of modulation signal, so they all contain the information of modulation signal. From the point of information transmission, one sideband can be further suppressed and only one sideband (upper sideband or lower sideband) can be retained. This not only further saves transmission power, but also reduces the bandwidth by half, which is beneficial for shortwave communication with extremely crowded channels.

This modulation that suppresses the carrier and transmits only one sideband is called SSB AM, and it is represented by SSB/SC-AM.

The modulation signal u_Ω and the carrier signal u_c are multiplied by a multiplier to obtain the DSB signal that suppresses the carrier, and then the SSB signal is obtained by filtering one sideband (upper sideband or lower sideband) from the DSB signal by a bandpass filter.

From eq. (4.14), we know that after passing the sideband filter, the upper or lower sideband signal can be obtained.

The upper sideband signal is

$$u_{\text{SSBH}} = \frac{1}{2} A U_{\Omega m} U_{cm} \cos(w_c + \Omega)t \tag{4.15}$$

The lower sideband signal is

$$u_{\text{SSBL}} = \frac{1}{2} A U_{\Omega m} U_{cm} \cos(w_c - \Omega)t \tag{4.16}$$

It is observed from the above two formulas that the amplitude of SSB signal is proportional to the amplitude $U_{\Omega m}$ of modulation signal, and its frequency varies with the frequency of modulation signal.

Figure 4.6 shows the time domain waveforms and spectrum diagrams of three kinds of amplitude-modulated signals under single tone and multitone modulation, respectively.

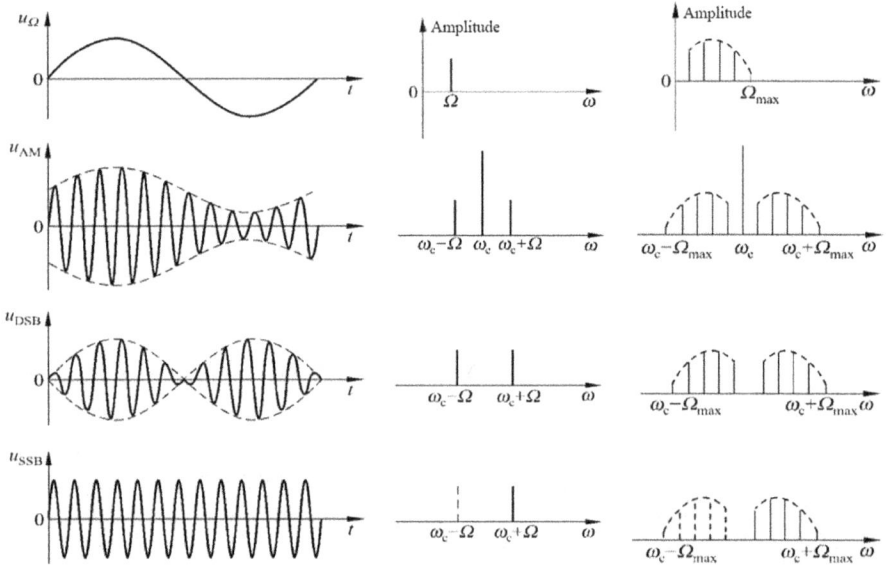

Fig. 4.6: Waveforms of three kinds of AM wave in time domain and frequency domain.

4.3 Amplitude-modulated circuits

In the radio transmitter, the method of AM is divided into two categories based on the level of power: high-level AM and low-level AM. The former produces amplitude-modulated wave that meets the requirement of output power directly at the last stage of transmitter, and the latter produces small power amplitude-modulated wave in the front stage of transmitter, and then amplifies it through linear power amplifier to reach the required transmission power level.

The standard AM wave is usually produced by the high-level amplitude-modulated circuit. Its advantage is that it does not need to adopt a low-efficiency linear amplifier, and the class C resonant power amplifier with high efficiency can be used to improve the efficiency of whole transmitter, but it must take into account the requirements of output power, efficiency and modulation linearity. The advantage of low-level amplitude-modulated circuit is that the power of modulator is small and the circuit is simple. Because of its small output power, it is often used in DSB modulation and low-level output system, such as signal generator.

4.3.1 High-level amplitude-modulated circuits

The high-level amplitude-modulated circuit is based on a class C-tuned power amplifier. In fact, it is a tuned power amplifier whose output voltage amplitude is controlled by the modulation signal [14].

Most modern, efficient, low-power, battery-powered wireless devices use class C amplifiers (or some other switching amplifiers) during the transmission phase. One of the main advantages of using class C amplifier in broadcast AM transmitter is that it only needs to be modulated at the last stage, and all the front stages can be driven at a constant signal level.

According to the different mode of modulation signal injection modulators, it can be divided into two types: base AM and collector AM.

4.3.1.1 Base amplitude-modulated circuit

Base amplitude-modulated circuit is shown in Fig. 4.7. It can be seen from the diagram that the high-frequency carrier signal u_ω is added to the base circuit of transistor through high-frequency transformer T_1, and the low FM signal u_Ω is added to the base circuit of transistor through low-frequency transformer T_2, and C_b is a high-frequency bypass capacitor, which is used to provide the path for the carrier signal.

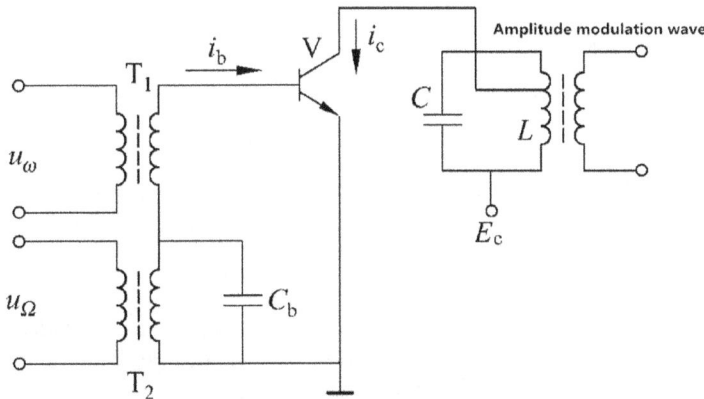

Fig. 4.7: Base amplitude-modulated circuit.

In the modulation process, the modulation signal u_Ω is equivalent to a slowly varying bias voltage, which makes the maximum value i_{cmax} of amplifier collector pulse current and the conduction angle θ vary according to the modulation signal. When u_Ω increases, i_{cmax} and θ increase, and when u_Ω decreases, i_{cmax} and θ decrease. Therefore, the amplitude of output voltage reflects the waveform of the modulation signal.

The waveform of transistor collector current i_c and the waveform of output voltage from the tuning loop are shown in Fig. 4.8. If the collector resonant circuit is tuned to the carrier frequency f_c, the output of amplifier will obtain the amplitude-modulated wave.

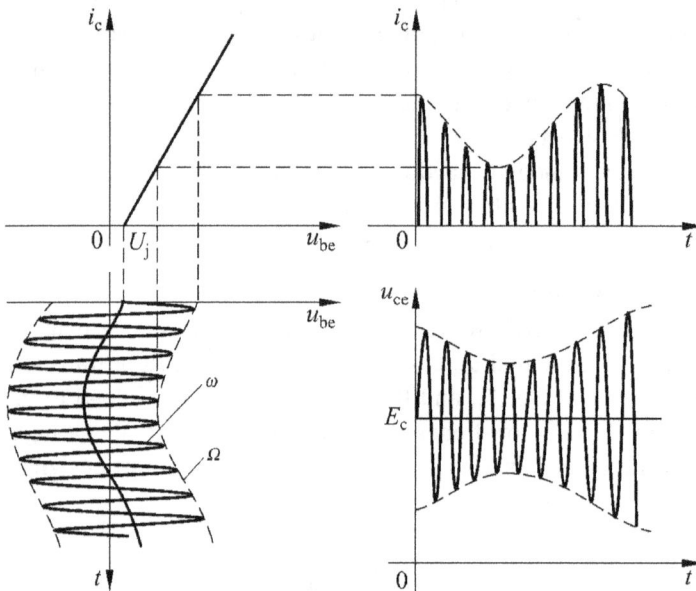

Fig. 4.8: Waveform of base amplitude modulation.

In Chapter 2, we studied the base modulation characteristics of class C power amplifier. From Fig. 2.26, we can see that the base modulation characteristic curve is nearly linear in the middle section, while the upper part and the lower part have great bending. The upper bending is the overvoltage of amplifier, and the lower bending is caused by the bending of starting part of transistor input characteristics.

In order to reduce the modulation distortion, the carrier operating point should be selected in the center of linear part of modulation characteristic so that the modulated amplifier always works in the undervoltage state in the range of modulation voltage. Larger AM index and better linear AM can be obtained.

It can be seen that the base amplitude-modulated circuit can only produce standard AM signals, and the class C power amplifier must work in the undervoltage state [15].

4.3.1.2 Collector amplitude-modulated circuit

Collector amplitude-modulated circuit is shown in Fig. 4.9. The high-frequency carrier signal is still added from the base, while the modulation signal u_Ω is added to the collector. R_1C_1 is the base self-supporting bias. Through the secondary transformer, the modulation signal u_Ω connected with E_c in series can be regarded as a slowly changing integrated power supply E_{cc} ($E_{cc} = E_c + u_\Omega$).

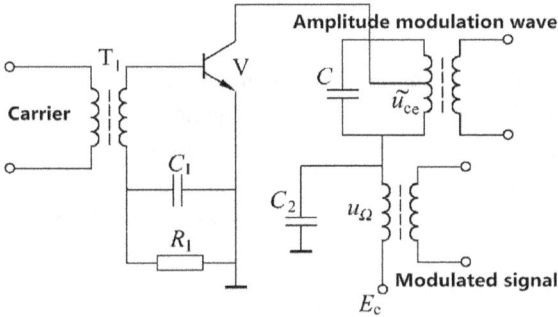

Fig. 4.9: Collector amplitude-modulated circuit.

In Chapter 2, we studied the collector modulation characteristics of class C power amplifier. From Fig. 2.25, we can see that the collector modulation characteristic curve is nearly linear in the middle section while the upper part and the lower part have great bending.

In order to reduce the modulation distortion, the carrier operating point should be selected in the center of linear part of modulation characteristic so that the modulated amplifier always works in the overvoltage state in the range of modulation voltage. Larger AM index and better linear AM can be obtained.

If the input signal is a high-frequency cosine carrier and the output LC loop is tuned to ω_c, then the output voltage is

$$u_{ce}(t) = kE_{cc} \cos \omega_c t = k[E_c + u_\Omega(t)] \cos \omega_c t \tag{4.17}$$

where k is a proportionality constant.

It can be seen that the collector amplitude-modulated circuit can only produce standard AM signals, and the class C power amplifier must work in the overvoltage state [16].

4.3.2 Low-level amplitude-modulated circuits

Low-level AM mainly produces low-power amplitude-modulated waves used in the front of transmitters. Analog multiplier is a common device in low-level amplitude-modulated circuits. It can not only realize standard AM wave, but also realize DSB amplitude-modulated wave and SSB amplitude-modulated wave.

One of the main disadvantages of low-level amplitude modulators is that a linear amplifier must be used after modulation. Linear amplifiers are relatively inefficient in power transmission, so these modulation schemes do not apply to high-power RF transmitters or modern battery-powered wireless devices on commercial radio stations.

In the following pages, we mainly introduce two commonly used circuits for low-level AM.

4.3.2.1 Integrated analog multiplier amplitude-modulated circuit

An integrated analog multiplier chip MC1596G is used to generate amplitude-modulated wave as shown in Fig. 4.10. Modulation signal is added from pin 1, carrier signal is added from pin 10, amplitude-modulated signal is output by pin 6 through capacitance 0.1 μF. The potentiometer 51 kΩ between pins 1 and 4 is used to adjust the AM factor m_a.

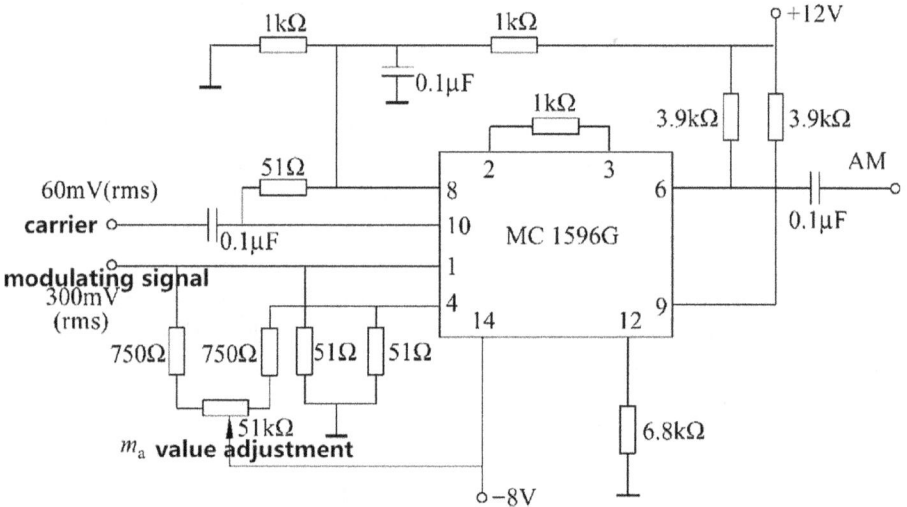

Fig. 4.10: Integrated analog multiplier amplitude-modulated circuit.

The circuit shown in Fig. 4.10 can also be used to form a DSB amplitude-modulated circuit. The difference is that adjusting the potentiometer makes the DC potential difference between pin 1 and pin 4 zero, that is, the input of pin 1 is only AC modulation signal.

The modulation signal and the carrier signal are multiplied by the integrated analog multiplier to obtain the DSB signal that suppresses the carrier, and then the SSB signal is obtained by filtering one sideband (upper sideband or lower sideband) from the DSB signal by the bandpass filter.

In order to reduce the current passing through the potentiometer and make it easy to set zero accurately, the resistance of two resistors 750 Ω can be increased, for example, by increasing the resistance to 10 kΩ each.

4.3.1.2 Diode ring amplitude-modulated circuit

Diode ring amplitude-modulated circuit, also known as diode double-balanced circuit, is mainly used to produce amplitude-modulated signal with suppressed carrier.

As shown in Fig. 4.11(a), the circuit consists of four diode rings. The carrier u_c is connected from the primary coil of transformer T_1, the modulation signal u_Ω is connected between the secondary midpoint of transformer T_1 and the midpoint of primary coil of T_2, and the secondary coil of transformer T_2 outputs the modulated signal. The equivalent circuit is shown in Fig. 4.11(b).

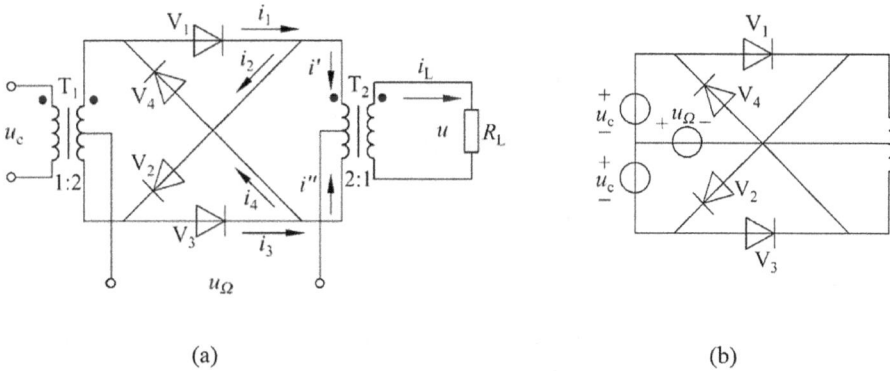

Fig. 4.11: Diode ring amplitude-modulated circuit, (a) schematic, (b) equivalent circuit.

Assume that the modulation signal be a monophonic cosine signal, that is

$$u_\Omega(t) = U_{\Omega m} \cos \Omega t \tag{4.18}$$

and carrier signal is

$$u_c(t) = U_{cm} \cos \omega_c t \tag{4.19}$$

Diode ring amplitude modulator can work in both small signal state and large signal state. In general, the amplitude of carrier signal is very large to control diodes that work in the switching state, and the amplitude of modulation signal is small.

When u_c is in the positive half cycle, V_1, V_2 are on, and V_3, V_4 are off; when u_c is in the negative half cycle, V_3, V_4 are on, and V_1 and V_2 are off. In order to analyze the current of diodes, the corresponding circuits are drawn in Fig. 4.11(b) as shown in Fig. 4.12.

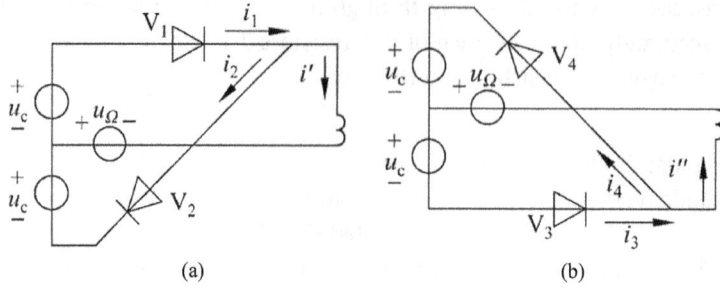

Fig. 4.12: Diode equivalent circuits, (a) $u_c > 0$, (b) $u_c < 0$.

Since diodes work in the switching state, V_1 and V_2 are on when $u_c > 0$. According to Fig. 4.12(a), ignore the load effect; hence, we can write the equations of current i_1 and i_2 as follows:

$$i_1 = g_d K_1(\omega_c t)(u_c + u_\Omega) \tag{4.20}$$

$$i_2 = g_d K_1(\omega_c t)(u_c - u_\Omega) \tag{4.21}$$

where g_d is the conductance of diode, and $K_1(\omega_c t)$ is unidirectional switching function.

According to the law of node current, we can obtain

$$i' = i_1 - i_2 = 2g_d K_1(\omega_c t)u_\Omega \tag{4.22}$$

Similarly, V_3 and V_4 are on when $u_c < 0$. According to Fig. 4.12(b), ignore the load effect; hence, we can write the equations of current i_3 and i_4 as follows:

$$i_3 = -g_d K_1(\omega_c t + \pi)(u_c - u_\Omega) \tag{4.23}$$

$$i_4 = -g_d K_1(\omega_c t + \pi)(u_c + u_\Omega) \tag{4.24}$$

It is worth noting that since V_3 and V_4 are on when u_c is in the negative half cycle, the corresponding switching function should be $K_1(\omega_c t + \pi)$.

According to the law of node current, we can obtain

$$i'' = i_3 - i_4 = 2g_d K_1(\omega_c t + \pi)u_\Omega \tag{4.25}$$

Therefore, we can write that the current that flows through the load R_L is

$$i_L = i' - i'' = 2g_d u_\Omega [K_1(\omega_c t) - K_1(\omega_c t + \pi)]$$

$$= 2g_d u_\Omega K_2(\omega_c t) \tag{4.26}$$

And the output voltage u at both ends of load R_L is

$$u = 2g_d R_L u_\Omega K_2(\omega_c t)$$

$$= 2g_d R_L U_{\Omega m} \cos \Omega t \cdot \frac{4}{\pi} \left[\cos \omega_c t - \frac{1}{3} \cos 3\omega_c t + \frac{1}{5} \cos 5\omega_c t + \cdots \right] \qquad (4.27)$$

It can be seen that the output voltage of diode ring modulator contains $(2n{-}1)\omega_c \pm \Omega$ side frequency components, but no carrier component, so if a bandpass filter (center frequency is ω_c, bandwidth $> 2\,\Omega$) is added to the output voltage of modulator, a DSB amplitude-modulated signal with the suppressed carrier can be obtained.

If the access positions of carrier and modulation signal are exchanged in Fig. 4.11, the multiplying effect can also be achieved. The analytical method is similar and no longer repeated.

From the above analysis, we can see that the center taps of transformers T_1 and T_2 must be strictly symmetrical, and the characteristics of four diodes should be the same; otherwise, the carrier wave cannot be suppressed, resulting in unwanted leakage.

In order to eliminate the asymmetry of circuit, the improved diode ring modulator circuit is shown in Fig. 4.13. It can be seen that the potentiometer W whose resistance is 50–100 Ω is added at the center tap of T_2, and the center point is symmetrical by adjusting W. In addition, four resistors $R_1 \sim R_4$ are inserted in the diode branch to reduce the asymmetry caused by the internal resistance inconsistency and instability of diode.

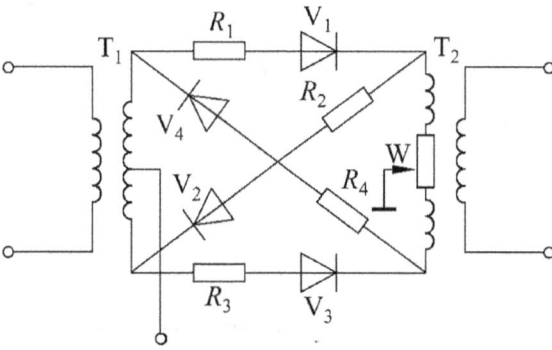

Fig. 4.13: The actual circuit of diode ring modulator.

4.4 Amplitude demodulation circuits

When the modulated signal reaches the receiving antenna, the receiver must extract the embedded information in some way and separate it from the high-frequency carrier signal. This information recovery process is called demodulation or detection. Amplitude demodulation is based on a basic mechanism similar to AM, in which a nonlinear device is used to multiply two waves and complete the frequency shift.

However, the demodulation process is centered on the carrier frequency w_c and the signal spectrum moves down to the baseband and returns to its initial position in the frequency domain. AM and demodulation need a frequency shift process, and both processes move the frequency spectrum at the distance w_c on the frequency axis. Both processes need a nonlinear circuit to complete the task.

Amplitude detection circuit includes two types: envelope detection circuit and synchronous detection circuit. The envelope detection circuit can only detect the standard amplitude-modulated signal, while the synchronous detection circuit can detect all kinds of amplitude-modulated signals.

In the envelope detection circuits, the diode large signal peak envelope detector is usually used, so it is necessary to avoid working in the small signal detection state, because the square law detector is realized when the small signal is input and the nonlinear distortion is serious.

4.4.1 Large signal peak envelope detection circuits

The schematic diagram of a diode large signal peak envelope detector is shown in Fig. 4.14, and the voltage amplitude of an input signal is generally larger than 500 mV, so the diode works in the switching state.

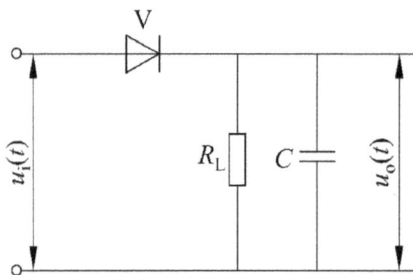

Fig. 4.14: Large signal peak envelope detector.

4.4.1.1 Working principle
We explain the working principle of a diode peak envelope detector by changing the waveform in the time domain, and Fig. 4.15 shows how it works.

Assume that the break-over voltage of diode is zero. When the input signal $u_i(t)$ is positive, the diode is switched on, the signal is charged through the diode to the capacitance C and the output voltage $u_o(t)$ increases with an increase in the charging voltage. When $u_i(t)$ decreases and is less than $u_o(t)$, the diode is reversed off and

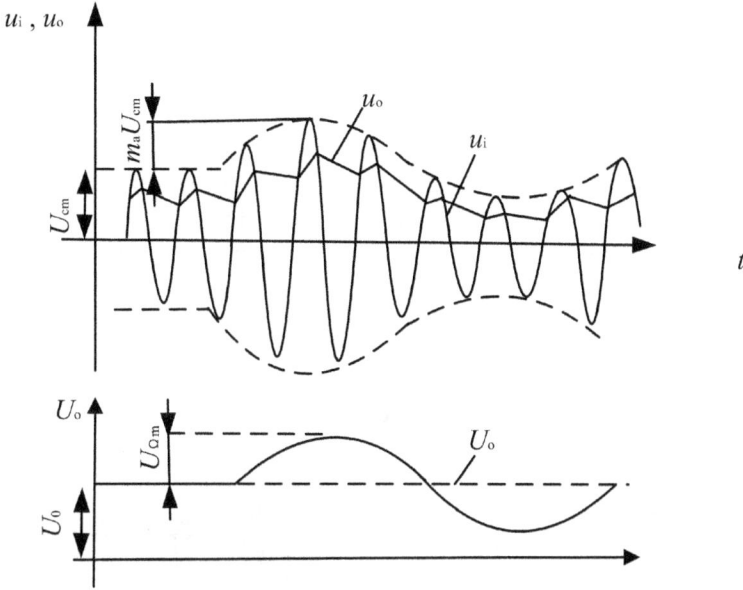

Fig. 4.15: Waveform of large signal peak envelope detector.

stops charging to the capacitor C, $u_o(t)$ discharges through the resistance R_L and $u_o(t)$ decreases with the discharge.

When charging, the positive on-resistance of diode is smaller, the charge is faster and $u_o(t)$ increases at a rate close to $u_i(t)$. During discharge, $u_o(t)$ fluctuates little because the resistance R_L is much larger than r_D (usually R_L = 5 –10 kΩ), the discharge is slower and $u_o(t)$ is almost close to the amplitude of $u_i(t)$.

If $u_i(t)$ is a high-frequency constant amplitude wave, $u_o(t)$ is a DC voltage U_O (ignoring a small amount of high-frequency components), which is exactly the rectifier circuit with filter capacitance.

When the amplitude of input signal $u_i(t)$ increases or decreases, the output voltage $u_o(t)$ of the detector will increase or decrease proportionally. When the input signal $u_i(t)$ is an amplitude-modulated wave, the output voltage $u_o(t)$ of detector changes with the envelope of amplitude-modulated wave; thus, the modulation signal is obtained and the detection effect is completed. Because the output voltage $u_o(t)$ is close to the peak value of input voltage $u_i(t)$, the detector is called the peak envelope detector.

4.4.1.2 Performance index

4.4.1.2.1 Detection efficiency

Detection efficiency, also known as voltage transmission coefficient, is expressed by η_d. It is one of the main performance indexes of detector. It is used to describe the ability of detector to convert high-frequency amplitude-modulated wave to low FM voltage.

The input and output waveforms of detector are shown in Fig. 4.16. When the input is a high-frequency constant amplitude wave, the output is a DC voltage (Fig. 4.16(a)), when the input is an amplitude-modulated wave, the output is a modulation signal (Fig. 4.16(b)).

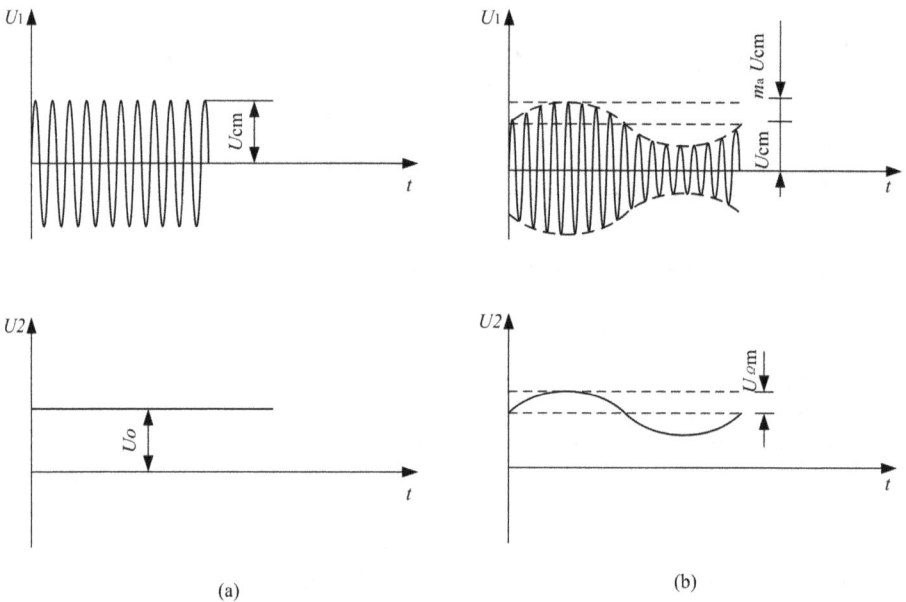

Fig. 4.16: Detector waveforms, (a) carrier waveforms, (b) AM waveforms.

If the amplitude of input constant voltage is U_{cm} and the output DC voltage of detector is U_0, the detection efficiency η_d is defined as

$$\eta_d = \frac{U_0}{U_{cm}} \tag{4.28}$$

For the amplitude-modulated wave, detection efficiency η_d is defined as the ratio of amplitude of output low-frequency voltage to amplitude of envelope of input high-frequency amplitude-modulated wave, that is

$$\eta_d = \frac{U_{\Omega m}}{m_a U_{cm}} \tag{4.29}$$

The higher the detection efficiency of detector, the larger the output low-frequency signal can be obtained for the same input signal. The detection efficiency of a general diode detector is always less than 1, and the circuit is designed as close to 1 as possible.

According to the analysis of detection principle mentioned earlier, as long as the charge is fast enough and the discharge is slow enough, the amplitude of output low-frequency voltage will approach the amplitude of envelope (peak value) of input signal, so the detection efficiency is close to 1. Specifically, the detection efficiency depends on the circuit parameters R_L, C, on-resistance r_d of diode and the amplitude of input signal.

Figure 4.17 is a set of measured curves of a large signal detector circuit. It can be seen from the diagram that the detection efficiency η_d increases with the increase of $\omega_c C R_L$ for a certain R_L. The larger $R_L C$ is, the slower the discharge is, and the larger ω_c is, the discharge time is shorter in one cycle. Both of them are conducive to the accumulation of more charge on C and increase the efficiency η_d.

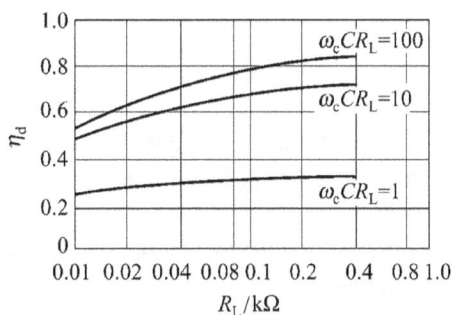

Fig. 4.17: Influence of detector parameters and frequency on η_d.

However, the effect of $\omega_c C R_L$ on η_d is uneven. As shown in the diagram, the effect of change of $\omega_c C R_L$ on η_d is very large when $\omega_c C R_L = 1\text{--}10$, and when $\omega_c C R_L = 10\text{--}100$, the effect on η_d is much smaller. When $\omega_c C R_L > 100$, the effect is basically insignificant. Therefore, when selecting $R_L C$ parameters, it is advisable to take a larger one, but not too large. Generally, we can choose $\omega_c C R_L = 10\text{--}100$, because too much is not obvious to improve η_d, but will cause distortion (will be discussed in the detection distortion).

When the positive resistance r_d of diode is small, the charging voltage is high, which is beneficial to improve η_d. When the reverse resistance of diode is small, a part of the charge will leak through the diode, which will reduce η_d. Therefore, in order to improve the detection efficiency η_d, it is advisable to select the diode with small forward resistance and large reverse resistance.

When the input signal is large, the forward resistance r_d of diode is small, so the detector efficiency η_d is improved.

4.4.1.2.2 Input resistance

Input resistance is another important performance index of envelope detector. As shown in Fig. 4.18, for the input high-frequency signal, the detector is equivalent to a load of the former intermediate-frequency (IF) amplifier, which is the equivalent input resistance R_{in} of detector.

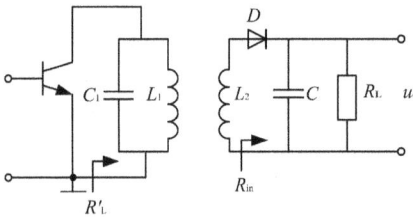

Fig. 4.18: The equivalent input resistance of envelope detector.

Assuming that the resonant resistance of LC loop of former IF amplifier is R_0, the equivalent load R_L' of former IF amplifier should be $R_0//R_{in}$. The larger the input resistance R_{in}, the smaller the effect on the previous stage circuit.

R_{in} is equal to the ratio of amplitude of input high-frequency voltage U_{cm} to the fundamental component of detection current I_{1m}, that is

$$R_{in} = \frac{U_{cm}}{I_{1m}} \tag{4.30}$$

When the current is cosine pulse, I_{1m} can be expressed by DC component I_0, and I_0 can be expressed by output voltage U_0 and detection resistor R_L, and for the large input signal, U_0 is very large and the angle of conduction θ is very small, $\alpha_1(\theta)/\alpha_0(\theta) \approx 2$, that is

$$R_{in} = \frac{U_{cm}}{I_{1m}} = \frac{U_{cm}}{\frac{\alpha_1(\theta)}{\alpha_0(\theta)} I_0} = \frac{U_{cm}}{\frac{\alpha_1(\theta)}{\alpha_0(\theta)} \frac{U_0}{R_L}} = \frac{U_{cm}}{\frac{\alpha_1(\theta)}{\alpha_0(\theta)} \frac{\eta_d U_{cm}}{R_L}} \approx \frac{R_L}{2\eta_d} \tag{4.31}$$

Equation (4.31) shows that when the detection efficiency η_d approaches 1, the input resistance R_{in} of large signal envelope detector is approximately equal to half of the load resistance R_L.

4.4.1.3 Detection distortion

There are two common distortion in large signal diode peak envelope detector: one is distortion caused by low discharge of filter capacitance, which is called diagonal distortion or inert distortion; the other distortion is caused by the DC voltage charged on the output coupling capacitor, which is called negative peak clipping distortion or bottom clipping distortion. The following are discussed separately.

4.4.1.3.1 Diagonal distortion

Referring to the circuit shown in Fig. 4.14, under normal conditions, the filter capacitor charges and discharges at high frequency once in a period, each time to be charged close to the peak voltage, so that the output of detector can basically keep up with the change of envelope. Its discharge is exponential and the time constant is $R_L C$.

Assuming that $R_L C$ is large, the discharge is very slow, and the envelope voltage may have decreased in several subsequent high-frequency cycles, but the voltage on C is still larger than the envelope voltage, which causes the diode to be off in reverse and lose the detection effect. The detection function will not be restored until the envelope voltage rises again to be larger than the voltage on C.

When the diode is off, the output voltage waveform of detector is the discharge waveform of C, which is inclined diagonal, as shown in Fig. 4.19, so it is called diagonal distortion, also called inert distortion. It is very obvious that the slower the discharge or the faster the envelope decline, the more likely this distortion will occur.

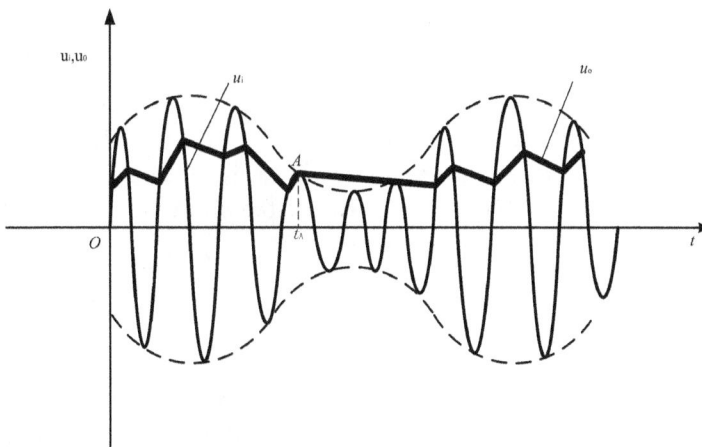

Fig. 4.19: Diagonal distortion waveform.

Next, we analyze how to avoid diagonal distortion. We assume that the input signal of detector is an AM wave modulated by a single-frequency cosine signal, and its envelope varies with time as a function of

$$u_m(t) = U_{cm}(1 + m_a \cos \Omega t)$$

At time t_A, the time function of capacitor C discharge is

$$u_c(t) = U_{cm}(1 + m_a \cos \Omega t_A)e^{-\frac{t-t_A}{R_L C}}$$

Because the diagonal distortion is caused by the fact that the speed of capacitance discharge is slower than the envelope descent speed of input signal, so the condition for nondiagonal distortion can be written out as follows:

Capacitor discharge speed $\geq$ envelope descent velocity

That is

$$\left|\frac{du_c(t)}{dt}\right|_{t=t_A} \geq \left|\frac{du_m(t)}{dt}\right|_{t=t_A} \tag{4.32}$$

where

$$\left|\frac{du_m(t)}{dt}\right|_{t=t_A} = U_{cm}m_a \Omega \sin \Omega t_A \tag{4.33}$$

$$\left|\frac{du_c(t)}{dt}\right|_{t=t_A} = \frac{U_{cm}}{R_L C}(1 + m_a \cos \Omega t_A) \tag{4.34}$$

Based on eqs. (4.32)–(4.34), we can get that the condition of no diagonal distortion is

$$R_L C \leq \frac{\sqrt{1 - m_a^2}}{m_a \Omega} \tag{4.35}$$

Equation (4.35) shows that the larger the AM coefficient m_a or the modulation frequency Ω is, the faster the envelope changes, which will easily cause diagonal distortion. The larger the discharge time constant $R_L C$ is, the slower the capacitor discharges, which will also easily cause diagonal distortion.

4.4.1.3.2 Negative peak clipping distortion

In the receiver, the output of detector is usually coupled to the next stage by a large capacitance C_1 (5 –10 μF). For the DC output of detector, there is a DC voltage U_0 ($U_0 = \eta_d U_{cm}$) on C_1 which can be regarded as a DC source. With the aid of theory of active two-terminal network, C_1, R_L, R_i (R_i is input resistance of next-stage low-frequency amplifier) in Fig. 4.20(a) can be replaced by the equivalent circuit as shown in Fig. 4.20.

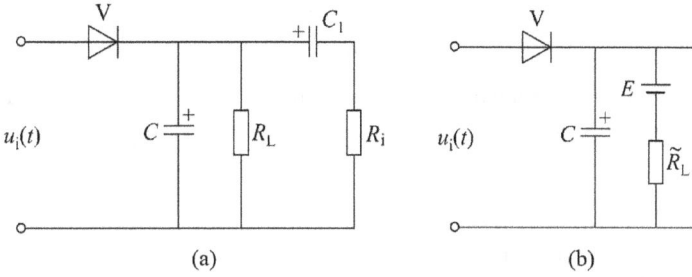

Fig. 4.20: Schematic diagram of negative peak clipping distortion, (a) schematic circuit, (b) equivalent circuit.

$$E = \frac{R_L}{R_L + R_i} \quad U_0 = \frac{R_L}{R_L + R_i} \eta_d U_{cm}, \quad \tilde{R}_L = R_L // R_i$$

If the modulation of input signal is so deep that the amplitude of input signal is smaller than E in $t_1 \sim t_2$ time, the diode will be in a reverse cut-off state during this period and the voltage on the capacitor C is equal to E, so the negative peak of output waveform is cut. Negative peak clipping distortion is generated, as shown in Fig. 4.21.

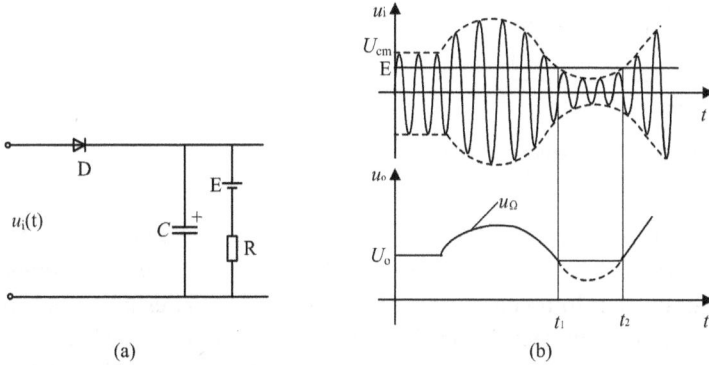

Fig. 4.21: Waveform of negative peak clipping distortion, (a) equivalent circuit, (b) input and output waveforms.

It can be seen that to avoid negative peak clipping distortion, the minimum of input signal $U_{cm} (1-m_a)$ is required to be greater than or equal to E, that is

$$(1 - m_a)U_{cm} \geq E = \frac{R_L}{R_L + R_i} \eta_d U_{cm}$$

$$\Rightarrow m_a \leq 1 - \eta_d \frac{R_L}{R_L + R_i} \tag{4.36}$$

For simplicity, if the detection efficiency η_d is equal to 1, then the condition that negative peak clipping distortion cannot be generated is

$$m_a \leq 1 - \frac{R_L}{R_L + R_i} = \frac{R_i}{R_L + R_i} = \frac{\tilde{R}_L}{R_L} \tag{4.37}$$

From eq. (4.37), it can be seen that the larger the modulation coefficient m_a or the smaller the ratio of AC load to DC load of detector, the more likely to produce negative peak clipping distortion.

In practical circuits, various measures can be taken to reduce the difference between AC load and DC load. For example, divide the R_L into R_{L1} and R_{L2}, and connect R_i to both ends of R_{L2} through a barrier capacitor C_1, as shown in Fig. 4.22.

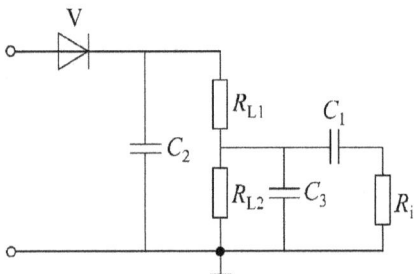

Fig. 4.22: Improved circuit of large signal envelope detector.

Figure 4.22 shows that the larger $R_L = R_{L1} + R_{L2}$ is, the smaller the difference of AC load and DC load resistance is, but the smaller the output audio voltage is. In order to resolve this contradiction in a compromise, $R_{L1}/R_{L2} = 0.1\sim0.2$ is usually used in practical circuits R_{L2} is also connected with the capacitor C_3 in order to further filtering out the high-frequency components and improve the high-frequency filter ability of detector.

4.4.2 Synchronous detection circuits

The envelope detector introduced in the previous section can only demodulate the standard AM wave, but not DSB and SSB signals, because the envelope of the latter two modulated signals does not reflect the changing law of modulation signal. Therefore, synchronous demodulation circuit must be used for demodulation of

DSB and SSB signals, and the most commonly used circuit is product synchronous demodulation circuit, and its block diagram is shown in Fig. 4.23.

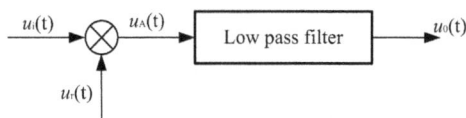

Fig. 4.23: Block diagram of product synchronous demodulator.

A product synchronous detector must provide a local carrier signal u_r at the receiving end, and it should be a synchronous signal with the same frequency and phase as the carrier signal at the transmitting end, which is the key to realize the synchronous demodulation. By multiplying the additional local carrier signal u_r with the input amplitude-modulated signal u_i from the receiver, the original modulation signal component and other harmonic components can be generated. After passing through the low-pass filter, the original modulation signal can be demodulated.

Set the input DSB signal and the synchronization signal as follows:

$$u_i = U_{im} \cos \Omega t \cos \omega_c t \tag{4.38}$$

$$u_r = U_{rm} \cos \omega_c t \tag{4.39}$$

Then the output voltage of multiplier is

$$u_A = A u_i u_r = A U_{im} U_{rm} \cos \Omega t \cos \omega_c t \cdot \cos \omega_c t$$

$$= \frac{1}{2} A U_{im} U_{rm} \cos \Omega t + \frac{1}{2} A U_{im} U_{rm} \cos \Omega t \cos 2\omega_c t \tag{4.40}$$

Obviously, the first item on the right-hand side of eq. (4.40) is the desired modulation signal, while the second term is a high-frequency component that can be filtered out by a low-pass filter.

Similarly, if the input signal is an SSB signal as follows

$$u_i = U_{im} \cos(\omega_c + \Omega)t \tag{4.41}$$

then the output voltage of multiplier is

$$u_A = A u_i u_r = A U_{im} U_{rm} \cos(\omega_c + \Omega)t \cdot \cos \omega_c t$$

$$= \frac{1}{2} A U_{im} U_{rm} [\cos \Omega t + \cos(2\omega_c + \Omega)t] \tag{4.42}$$

The output low FM signal can be obtained by filtering the high-frequency component with low-pass filter.

The product synchronous detector can not only demodulate DSB signals and SSB signals, but also demodulate standard AM signals. Assuming the input signal is

$$u_i = U_{cm}(1 + m_a \cos \Omega t) \cos \omega_c t \tag{4.43}$$

then the output voltage of multiplier is

$$u_A = Au_iu_r = AU_{cm}(1 + m_a \cos \Omega t) \cos \omega_c t U_{rm} \cos \omega_c t$$

$$= AU_{cm}U_{rm}(1 + m_a \cos \Omega t)\cos^2 \omega_c t$$

$$= \frac{AU_{cm}U_{rm}}{2} + \frac{AU_{cm}U_{rm}}{2} m_a \cos \Omega t + \frac{AU_{cm}U_{rm}}{2} \cos 2\omega_c t$$

$$+ \frac{AU_{cm}U_{rm}}{4} m_a \cos(2\omega_c + \Omega)t + \frac{AU_{cm}U_{rm}}{4} m_a \cos(2\omega_c - \Omega)t \tag{4.44}$$

Obviously, the second item on the right-hand side of eq. (4.44) is the desired modulation signal, while the high-frequency components can be filtered out by a low-pass filter.

From the above analysis, if we want the low FM signal to be demodulated without distortion, the synchronization signal and the carrier signal must be strictly in the same frequency and in the same phase; otherwise, the demodulation distortion will be caused.

For example, if u_r is not synchronized that has frequency difference $\Delta\omega$ and phase difference $\Delta\varphi$ as follows

$$u_r = U_{rm} \cos[(\omega_c + \Delta\omega)t + \Delta\varphi] \tag{4.45}$$

then the output demodulated signal is

$$u_0 = \frac{1}{2}AU_{cm}U_{rm}m_a \cos(\Delta\omega t + \Delta\varphi) \cos \Omega t \tag{4.46}$$

From eq. (4.46), it is very obvious that the demodulation distortion is generated, and at the receiving end there will be a demodulated signal with a slow variation of strength, commonly known as the beat phenomenon. Therefore, keeping two carriers in the same frequency and phase is the key to synchronous detector without distortion.

Multiplier in product detector can be realized by the nonlinear device. The circuits previously involved in low-level AM can be used as a multiplier. For example, a product detector can be implemented using a diode ring multiplier, as shown in Fig. 4.24. A diode ring multiplier can be used for both AM and amplitude demodulation.

A product synchronous detector circuit using an integrated analog multiplier is shown in Fig. 4.25. The amplitude-modulated signal u_i (which may be AM wave, DSB wave or SSB wave) is input from pins 1 and 4 after coupling capacitance, and

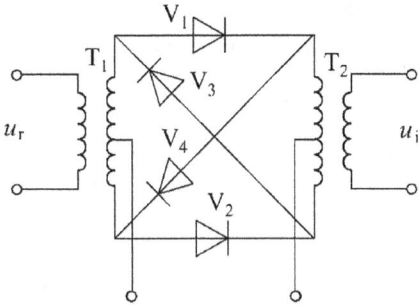

Fig. 4.24: Synchronous detector with diode ring multiplier.

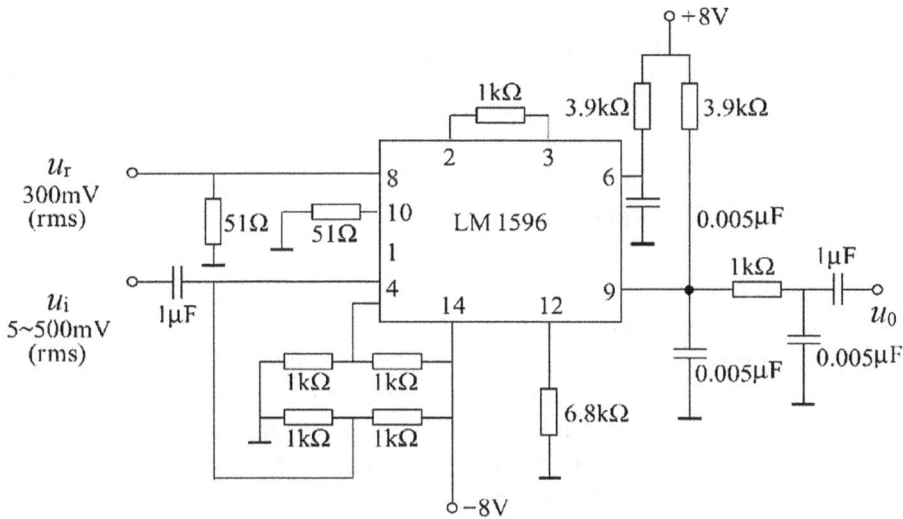

Fig. 4.25: Synchronous detector with an integrated analog multiplier.

the synchronization signal u_r is input from pins 8 and 10. The modulation signal u_o is extracted by an RC π-type low-pass filter after the single-end output of pin 9.

4.5 Mixers

In communications technology, it is often necessary to convert a signal from one frequency to another, usually by converting one modulated high-frequency signal into another of its kind at a lower frequency. The circuit that completes this frequency conversion is called a mixer.

The mixer is used for frequency conversion and is the key component of modern radio frequency system. A mixer converts radio frequency power from one frequency to another to make signal processing easier and cheaper. For example, in a superheterodyne medium-wave broadcast radio, the high-frequency signal received by the antenna (the carrier frequency of standard AM signal of stations is in 535 –1,605 kHz medium-wave band) is converted by the mixer to the IF signal of 465 kHz; and in the superheterodyne FM broadcasting radio, the signals with carrier frequency located in 88 –108 MHz are converted into the IF signals with 10.7 MHz; and the TV station signals located in the frequency range of 40 MHz to nearly 1,000 MHz are converted into video signals with an IF of 38 MHz.

The performance of receiver will be improved by using the mixer. This is due to:
(1) The mixer converts the high-frequency signal to the IF signal, amplifies the signal on the IF and the gain of amplifier can be very high but not self-excited, which helps to improve the sensitivity of receiver.
(2) The frequency received by the receiver is variable, but the IF after mixing is fixed, so the circuit structure can be simplified.
(3) The receiver is required to have good selectivity in a wide range of frequencies, which is difficult to do, but it can be done well for a fixed IF.

4.5.1 The basic principle of mixer

An electronic circuit that can multiply two AC signals is called a mixer. A mixer in a radio frequency system always refers to a circuit with nonlinear components. For two input monophonic signals w_1 and w_2, the output monophonic signal is the sum of input frequencies (i.e., $w_1 + w_2$) and the difference of input frequencies (i.e., $|w_1 - w_2|$). Note that in an audio system, the operator refers to the "mixing" of two channels, which is not in the sense of radio frequency: the sum and difference of input frequencies are not generated and the nonlinear component is not involved in the circuit. Instead, it is a linear addition of two signals so that two channels can be heard at the same time. The symbolic representation of these two operations shows the difference between linear addition and mixing of two AC signals.

Because the ideal linear time-invariant system cannot produce the output signal with new spectral components different from the input signal, in order to provide frequency conversion, the mixer must be nonlinear or time varying. Of course, some nonlinear treatments are better than others, so we only focus on the actual radio frequency mixer type.

The composition of the mixer is shown in Fig. 4.26. It shows that the input high-frequency-modulated signal induced by the receiving antenna is amplified by a low-noise amplifier (LNA) and added to the mixer together with the local oscillation signal (usually shows a high-frequency constant amplitude signal, and its frequency is f_L). After implementation of frequency conversion by multiplication of nonlinear devices, the IF-modulated signal is output through IF filter. The modulation

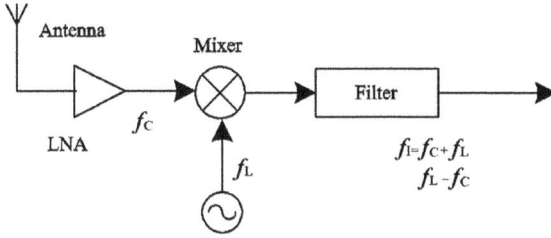

Fig. 4.26: Block diagram of mixer composition.

law between the output IF signal and the input RF signal is invariant, and the only difference is that the carrier frequency has been changed from f_c to f_I.

Therefore, a mixer consists of three parts: (1) nonlinear elements such as diodes, transistors, FETs and analog multipliers; (2) oscillators that produce local oscillations, commonly known as local oscillators; and (3) IF filter.

The core of modern mixer is the multiplication of two sine signals in the time domain. The basic usefulness of multiplication can be understood from the basic triangular identity, which shows that the multiplication of two sine wave functions produces two new sine waves.

IF frequency f_I can be the sum of f_c and f_L, or the difference between them. The former is called up-mixing, and the latter is called down-mixing. Down-mixing is usually used in receivers such as radio and television. IF filters can be used to select the sum or difference components that we need, that is, IF components.

The function of mixing is to shift the signal frequency from high frequency f_c to IF f_I, which is also a linear shift process of frequency spectrum. The spectrum structure before and after mixing has not changed, as shown in Fig. 4.27.

Figure. 4.27 shows that after mixing, the original input high-frequency AM signal is converted into IF AM signal at the output end. Compared with the two signals, the frequency of AM signal is only shifted from the high-frequency position to the IF position, and the relative magnitude of each spectrum component and the distance between them are consistent.

The following is a further mathematical analysis of mixing principle:

As we have learned earlier, in order to realize the linear shift of spectrum, we must realize the multiplication of two signals in the time domain. Assume that the input high-frequency amplitude-modulated signal u_i and the local oscillation signal u_L are, respectively:

$$u_i(t) = U_{cm}(1 + m_a \cos \Omega t) \cos \omega_c t \qquad (4.47)$$

$$u_L(t) = U_{Lm} \cos \omega_L t \qquad (4.48)$$

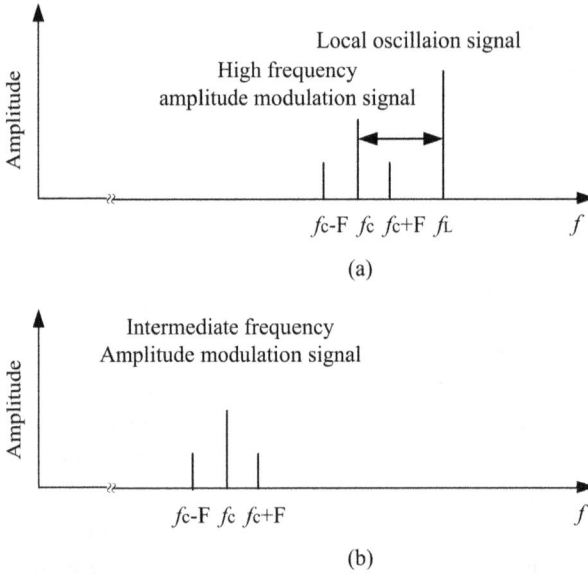

Fig. 4.27: Spectrum diagram of mixer, (a) before mixing, (b) after mixing.

They multiply to get

$$u_0(t) = U_{cm}U_{Lm}(1 + m_a \cos \Omega t) \cos \omega_c t \cos \omega_L t$$

$$= \frac{1}{2} U_{cm}U_{Lm}(1 + m_a \cos \Omega t)[\cos(\omega_c + \omega_L)t + \cos(\omega_L - \omega_c)t] \tag{4.49}$$

We can see from eq. (4.49) that the sum frequency $\omega_L + \omega_c$ and the difference frequency $\omega_L - \omega_c$ are generated by multiplying. If the difference frequency component $\omega_L - \omega_c$ is the required IF component ω_I, the difference frequency component can be obtained by the IF filter and the sum frequency component can be filtered out, that is

$$u_I(t) = \frac{1}{2} U_{cm}U_{Lm}(1 + m_a \cos \Omega t) \cos \omega_I t \tag{4.50}$$

We find that the IF signal after mixing is still AM signal.

4.5.2 The key technical indexes of mixer

The main technical parameters of a mixer include mixing gain, selectivity, working stability, nonlinear distortion, noise coefficient and isolation.

4.5.2.1 Mixing gain

Mixing gain is defined as the ratio of output IF signal to the input signal of the mixer. There are two types of voltage gain A_{uc} and power gain A_{pc}, usually expressed in decibels:

$$A_{UC} = 20\lg\frac{U_I}{U_s}\,(\text{dB}) \tag{4.51}$$

$$A_{PC} = 10\lg\frac{P_I}{P_s}\,(\text{dB}) \tag{4.52}$$

For receivers, A_{uc} (or A_{pc}) is large and can be used to improve sensitivity. Usually in broadcasting radios, A_{pc} is 20–30 dB, and in television receivers, A_{uc} is 6–8 dB.

4.5.2.2 Selectivity

In addition to producing useful IF signals, mixers also produce many unnecessary frequency terms. In order to make that the output of the mixer contains only the desired IF signals and suppresses the interference of other unnecessary frequencies, the output loop should have good selectivity. Frequency selection network or filter with high-quality factor Q can be adopted.

4.5.2.3 Working stability

In order to ensure the accuracy of IF produced by mixing, the frequency stability of a local oscillator signal is required to be high, so some measures such as frequency stabilization should be adopted.

4.5.2.4 Nonlinear distortion

Because the mixer operates in a nonlinear state, when the desired IF signal is obtained at the output end, there will be many unnecessary other frequency components, some of which will fall within the pass band of IF amplifier. The information contained in the output IF signal is different from that of the input signal, resulting in nonlinear distortion. In addition, in the mixing process, combined frequency interference and cross-modulation interference will be produced, which will affect the normal communication. Therefore, the distortion and interference should be minimized when designing and adjusting the mixing circuit.

4.5.2.5 Noise figure

The noise figure (NF) of the mixer is defined as the ratio of signal-to-noise power ratio at the input end to signal-to-noise power ratio at the output end of the mixer, which is usually expressed in decibels, that is

$$\mathrm{NF} = 10\lg \frac{(P_\mathrm{s}/P_\mathrm{n})_\mathrm{i}}{(P_\mathrm{l}/P_\mathrm{n})_\mathrm{o}} \, (\mathrm{db}) \tag{4.53}$$

Because the mixer is located at the front end of receiver, the noise generated by the mixer has a greatest influence on the whole receiver, so the smaller the noise coefficient of mixer is, the better. Obviously, in the ideal noise-free case, NF = 0 dB.

4.5.2.6 Isolation

Isolation refers to the amount of leakage or passage between the mixer ports, that is, the ratio of signal power of this port to the power it leaks to another port. Clearly, the greater the isolation, the better. The leakage of local oscillator signal is more important because of the large power of local oscillator.

4.5.3 The mixer circuits

There are many kinds of mixer circuits. Nonlinear devices such as diodes, transistors, FET and integrated analog multipliers can realize mixing. The transistor mixer has the advantages of high gain and low noise, but the frequency mixing distortion is large and the local oscillator leakage is serious. Because of its square law characteristic, FET mixer has smaller mixing distortion and larger dynamic range, but its mixing gain is smaller than that of transistor mixer. The diode ring mixer has a simple structure, low noise, low mixing distortion, large dynamic range and high working frequency (up to 1,000 MHz), and its disadvantage is small gain. The mixer circuit composed of integrated analog multiplier not only has small mixing interference, but is also easy to adjust, and the dynamic range of input signal is large.

4.5.3.1 Transistor mixer circuits

According to the different local oscillator signal access, in general, there are four circuit forms of transistor mixer. Figure 4.28(a), (b) shows two forms of common emitter circuits and Fig. 4.28(c), (d) shows two forms of common base circuits.

The common emitter circuit is mostly used in the case of lower frequency. In Fig. 4.28(a), input signal u_s and local oscillation signal u_L are injected by the base and the emitter, respectively. The mutual influence between u_s and u_L is small but the power required by the local oscillator is high. In Fig. 4.28(b), the input signal u_s and local oscillation signal u_L are injected by the base, which has a great influence on each other, but the power required by the local oscillator is small.

The common base circuit is mostly used in the case of higher frequency. When the frequency is not high, the mixing gain is lower than that of the common emitter circuit. Figure 4.28(d) has greater interaction than Fig. 4.28(c).

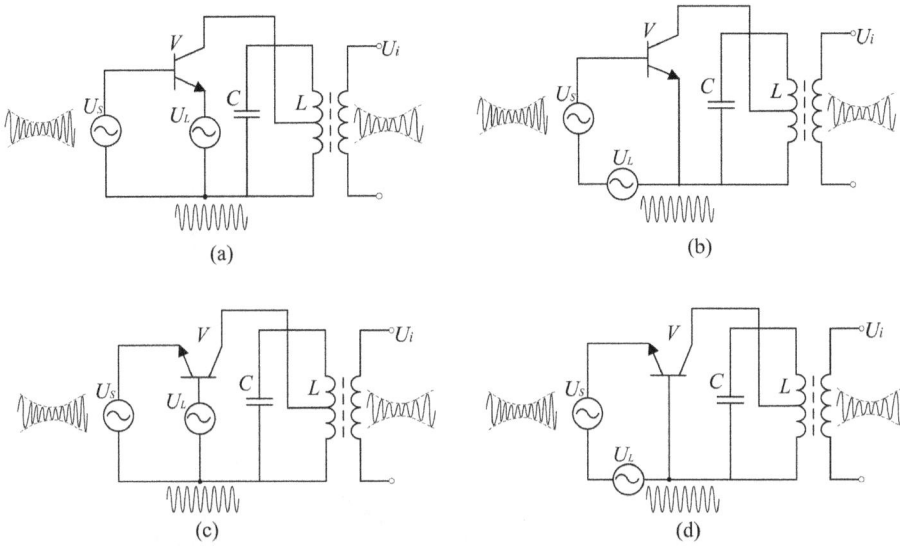

Fig. 4.28: Several forms of transistor mixer circuit, (a) common emitter with high UL, (b) common emitter with low UL, (c) common base with low UL, (d) common base with high UL.

The common feature of these circuits is that, no matter how the local oscillation voltage is injected, the input signal and the local oscillation signal are actually added between the base and emitter, and the frequency conversion is realized by using the nonlinear transfer characteristics of a transistor.

Figure 4.29 shows the transistor mixer circuit used in medium-wave broadcast radio. The local oscillator voltage is generated by the inductance feedback oscillator composed of V_2, which is added by the coupling coil L_c to the emitter of mixing transistor V_1. The input signal that is induced by the receiving antenna is added to the input loop through the coupling coil L_a, and then it is applied to the base of V_1 through the coupling coil L_b. The selection of input signal frequency and the corresponding local oscillator frequency is obtained by the variable capacitance of adjusting linkage, and the output is the IF signal whose frequency is 465 kHz.

In the actual circuit, the values of L_a and L_b are small, so for the frequency of input signal, the local oscillator circuit is seriously detuned, and its impedance at both ends of L_c is very small, which can be regarded as a short circuit; similarly, for the frequency of local oscillation signal, the input signal circuit is seriously detuned, and its impedance at both ends of L_b is very small and can be regarded as a short circuit. Therefore, both the input signal and the local oscillation signal have a good path, which can be effectively applied to the emitter junction of V_1. At the same time, it also effectively overcomes the leakage of local oscillation voltage to the antenna through the input signal loop and generates reverse radiation.

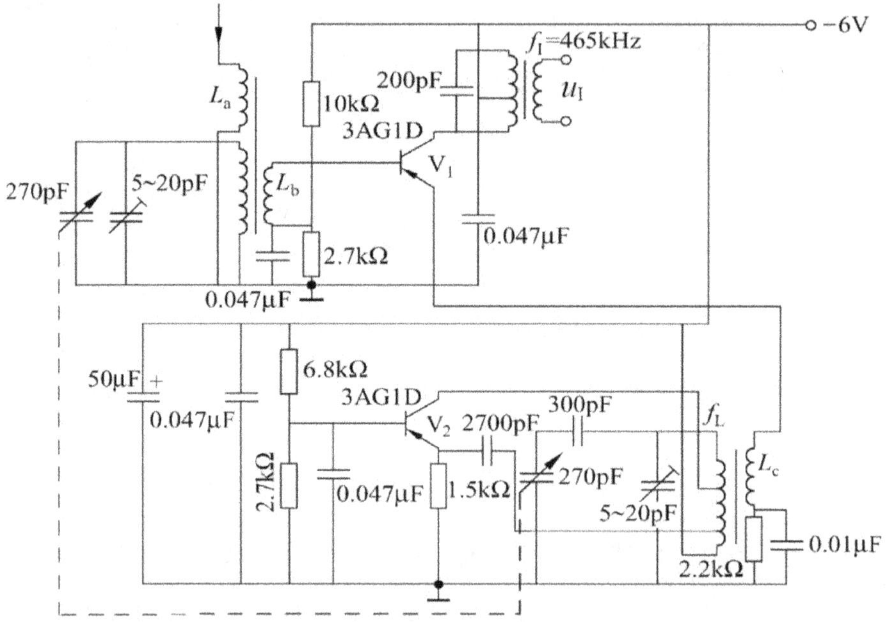

Fig. 4.29: Typical transistor mixer circuit in radio.

4.5.3.2 Diode ring mixer circuit

For many years, the diode ring mixer is the most common mixer topology for high-performance applications. The diodes provide the necessary switching action, which may be silicon junction, silicon Schottky barrier or gallium arsenide type. In the mixer whose practical operating frequency is more than dozens of MHz, the diode double-balanced mixer, also known as the diode ring mixer, is widely used as shown in Fig. 4.30.

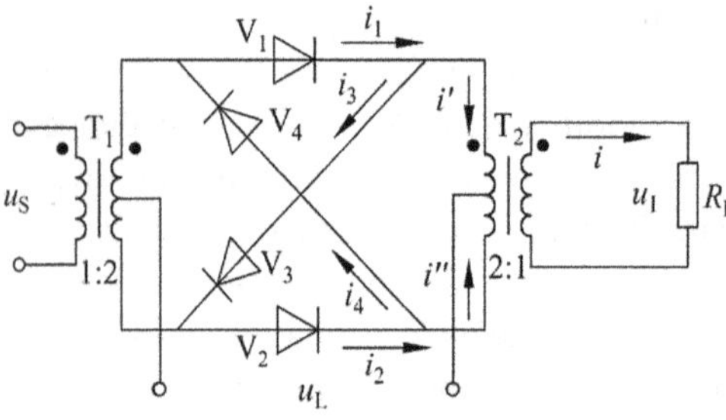

Fig. 4.30: Diode ring mixer circuit.

The input signal u_s ($u_s = U_{sm}\cos\omega_c t$) of mixer is usually small. When the local oscillation signal u_L ($u_L = U_{Lm}\cos\omega_L t$) is large enough, the diodes operate in the switching state controlled by u_L. When the local oscillation signal u_L is in the positive half cycle, the diodes V_1 and V_2 are on, V_3 and V_4 are off, and when the local oscillation signal voltage u_L is in the negative half cycle, the diodes V_3 and V_4 are on, V_1 and V_2 are off.

Relative to the local oscillation signal, the conduction polarity of V_1 and V_2 is opposite to that of V_3 and V_4. If the switching functions of V_1 and V_2 are $k_1(\omega_L t)$, then the switching functions of V_3 and V_4 are $k_1(\omega_L t + \pi)$.

According to Fig. 4.30, ignoring the load effect, we can write the equations of current i_1 and i_2 as follows:

$$i_1 = g_d K_1(\omega_L t)(u_L + u_s) \tag{4.54}$$

$$i_2 = g_d K_1(\omega_L t)(u_L - u_s) \tag{4.55}$$

where g_d is the conductance of diode, and $K_1(\omega_L t)$ is a unidirectional switching function.

The output current of a single balanced mixer consisting of V_1 and V_2 is

$$i' = i_1 - i_2 = 2g_d K_1(\omega_L t)u_s \tag{4.56}$$

Similarly, we can write the equations of current i_3 and i_4:

$$i_3 = -g_d K_1(\omega_L t + \pi)(u_L - u_s) \tag{4.57}$$

$$i_4 = -g_d K_1(\omega_L t + \pi)(u_L + u_s) \tag{4.58}$$

The output current of a single balanced mixer consisting of V_3 and V_4 is

$$i'' = i_3 - i_4 = 2g_d K_1(\omega_L t + \pi)u_s \tag{4.59}$$

So, we can write the current that flows through the load R_L is

$$i = i' - i'' = 2g_d u_s[K_1(\omega_L t) - K_1(\omega_L t + \pi)]$$
$$= 2g_d u_s K_2(\omega_L t) \tag{4.60}$$

And the output voltage u_I at both ends of load R_L is

$$u_I = 2g_d R_L u_s K_2(\omega_L t)$$
$$= 2g_d R_L U_{sm} \cos\omega_c t \cdot \frac{4}{\pi}\left[\cos\omega_L t - \frac{1}{3}\cos 3\omega_L t + \frac{1}{5}\cos 5\omega_L t + \cdots\right] \tag{4.61}$$

It can be seen that in the diode ring mixer circuit, as long as the circuit is symmetrical, the output voltage only contains $(2n-1)\omega_L \pm \omega_c$ frequency components, so if a bandpass filter (center frequency is $\omega_I = \omega_L - \omega_c$) is added to the output of the mixer, the IF voltage can be obtained and the mixing function is realized.

Before the analog multiplier came out, the diode ring mixer was a widely used circuit. Because the upper working frequency of circuit is high, the analog multiplier cannot replace the diode ring mixer in the frequency range above dozens of MHz. The diode ring mixer now available on the market is made of four diodes into integrated circuits.

4.5.3.3 Integrated analog multiplier mixer circuit

Figure 4.31 is a double-balanced mixer composed of MC1596G with a wideband input, output IF is tuned at 9 MHz, loop bandwidth is 450 kHz and the input level of a local oscillator is 100 mV. For 30 MHz of an input signal and 39 MHz of a local oscillation signal, mixing gain is 13 dB. When the output SNR is 10 dB, the sensitivity of input signal is 7.5 μV.

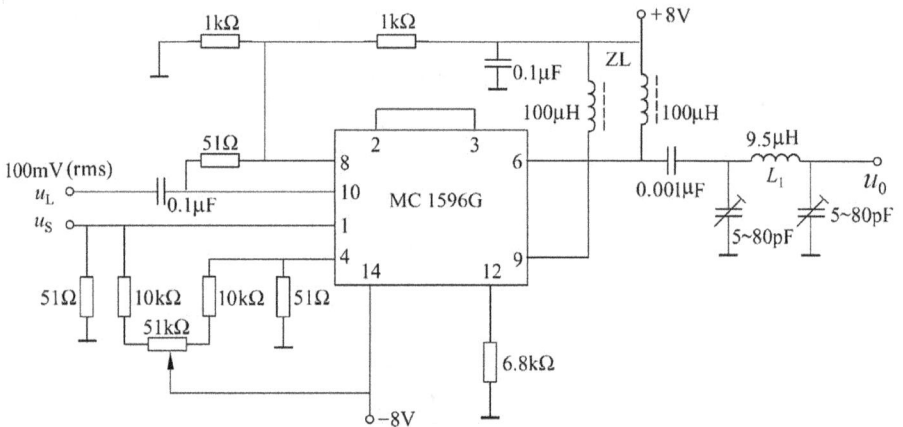

Fig. 4.31: The mixer circuit with MC1596G.

In addition to the use of integrated analog multipliers to achieve mixing, other devices with multiplicative characteristics can also be used to replace the analog multiplicators in Fig. 4.31. With the development of large-scale integrated circuit, mixing circuit has been integrated into special chip as a unit circuit.

4.5.4 Mixing interference and suppression methods

In general, due to the nonlinearity of mixer, the mixer will produce various kinds of interference and distortion, including interference whistle, side-channel interference, cross-modulation interference and intermodulation interference. The following is a discussion of the causes and suppression methods of common interference.

4.5.4.1 Combined frequency interference (interference whistle)

Interference whistle is a combined frequency interference produced by the useful signal and the local oscillating signal. When the input ends of a mixer are simultaneously added to the useful signal whose frequency is f_c and the local oscillating signal whose frequency is f_L, in general, because of the nonlinearity of a mixer, in the output current of mixer, there will be a large number of combined frequency components as follows:

$$f_{p,q} = |\pm pf_L \pm qf_c| \tag{4.62}$$

Their amplitudes decrease rapidly with the increase of $(p + q)$. Only one transform channel $(p = q = 1)$ is useful, which can convert the input signal frequency to the desired IF (such as $f_L - f_c = f_I$), while the other large number of transformation channels are useless, and some are even very harmful.

For some p, q values (except $p = q = 1$), if $f_{p,q}$ is very close to the IF, that is

$$|\pm pf_L \pm qf_c| \approx f_I \pm F \tag{4.63}$$

where F is the audio frequency; then the input signal can be converted to an IF signal through the useful channel of $p = q = 1$, but also can be converted to a parasitic signal whose frequency is close to the IF frequency through those channels of p, q that satisfy eq. (4.63).

For example, when $f_c = 931$ kHz, $f_L = 1{,}396$ kHz, $f_I = 465$ kHz, the combined frequency component corresponding to $p = 1$, $q = 2$ is $1{,}396{-}2 \times 931 = 466$ kHz $= 465 + 1$ kHz. The useless frequency component of 466 kHz is located near the IF and can be passed through the IF amplifier. In this way, the listener will hear both the useful signal and the whistle sound formed by the beat signal $(F = 1$ kHz$)$ detected by the detector. Therefore, the combined frequency interference is also called interference whistle.

Observe eq. (4.63) satisfying interference whistle, which can be decomposed into the following four expressions:

$$\begin{cases} pf_L - qf_c = f_I \pm F \\ qf_c - pf_L = f_I \pm F \\ pf_L + qf_c = f_I \pm F \\ -qf_c - pf_L = f_I \pm F \end{cases}$$

If $f_L - f_c = f_I$, then only the first two of above four forms are likely to be valid, and the latter two are invalid. By combining the first two forms, the frequency of the input useful signal which can cause interference whistle is obtained as follows:

$$f_c \approx \frac{p \pm 1}{q-p} f_I \tag{4.64}$$

Equation (4.64) shows that if p and q are different positive integers, there are infinite frequencies of input useful signals, which will produce interference whistles, and their values are close to the integral or fractional times of f_I.

In fact, the receiving band of any receiver is limited, only those falling in the receiver band will produce interference whistles. Moreover, because the amplitude of the combined frequency component always decreases rapidly with the increase of $(p + q)$, only the input useful signal corresponding to the lower value of $(p + q)$ can produce obvious interference whistles. However, when the value of $(p + q)$ is higher, the interference whistles can be ignored.

It can be seen that as long as the signal frequency which produces the strongest interference whistle is moved beyond the receiving band, the harmful effects of interference whistles can be greatly reduced. From eq. (4.64), we can see that the interference whistle of $p = 0$, $q = 1$ is the strongest, and the corresponding input signal frequency is close to the IF, that is $f_c \approx f_I$. Therefore, in order to avoid the strongest interference whistle, the IF frequency of receiver is always selected outside the receiving band. For example, the receiving band of medium-wave radio is 535 –1,605 kHz, and its IF frequency f_I is defined as 465 kHz.

4.5.4.2 Side-channel interference

Side-channel interference is a combined frequency interference produced by the external interference signal and the local oscillating signal. It seems to bypass the main channel f_c and enter the IF circuit through another channel, so the interference is called side-channel interference, and also called the parasitic channel interference.

When the input ends of the mixer are simultaneously added to the external interference signal whose frequency is f_M and the local oscillating signal whose frequency is f_L, in general, because of the nonlinearity of mixer, in the output current of mixer, there will be a large number of combined frequency components as follows:

$$f_{p,q} = |\pm p f_L \pm q f_M| \tag{4.65}$$

If for some p, q values, $f_{p,q}$ is very close to the IF, that is

$$|\pm p f_L \pm q f_M| = f_I \tag{4.66}$$

then the interference signals can be converted from f_M to f_I through these channels so that they can pass through the IF amplifier smoothly. Therefore, the listener can hear the voice of the interference signal.

Due to the limitation of $f_L - f_c = f_I$, eq. (4.66) that produces side-channel interference can only be established by the following two expressions:

$$\begin{cases} pf_L - qf_M = f_I \\ qf_M - pf_L = f_I \end{cases}$$

By combining them, the input interference frequency of side-channel interference can be obtained as follows:

$$f_M = \frac{p}{q} f_L \pm \frac{1}{q} f_I = \frac{p}{q} f_c + \frac{p \pm 1}{q} f_I \tag{4.67}$$

Theoretically, there are infinite number of f_M, which can form side-channel interference. In fact, only the external interference signal corresponding to the lower value of $(p + q)$ can form stronger side-channel interference, while corresponding to the higher value of $(p + q)$, side-channel interference can generally be ignored.

Based on eq. (4.67), two frequencies forming the strongest side-channel interference can be obtained. One is the side channel for $p = 0$, $q = 1$ and correspondingly $f_M = f_I$. This is called IF interference. For such interference signals, mixers actually act as IF amplifiers. The other is the side channel for $p = 1$, $q = 1$ and correspondingly $f_M = f_L + f_I = f_c + 2 f_I$. If you think of f_L as a mirror, then f_M is the mirror of f_c, as shown in Fig. 4.32; therefore, it is called mirror frequency interference.

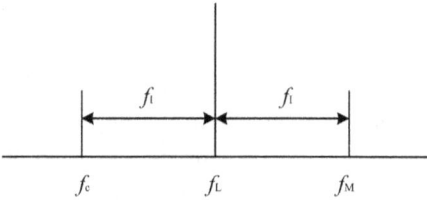

Fig. 4.32: Mirror frequency interference diagram.

If we rewrite eq. (4.67) to

$$f_c = \frac{q}{p} f_M - \frac{p \pm 1}{p} f_I \tag{4.68}$$

then it can also be explained: when f_M is fixed, the receiver can hear the sound of the interference signal on f_c. For example, when the input end of the mixer is acted on the interference signal of $f_M = 1{,}000$ kHz, based on eq. (4.68), we can find that the receiver can receive the sound of an interference signal at frequencies such as 1,070 kHz ($p = 1$, $q = 2$) and 767.5 kHz ($p = 2$, $q = 2$).

In order to suppress the side-channel interference, the frequency interval between the interference signal and the input useful signal must be increased, so that the interference signal can be filtered out by the filter before mixer, so that they are not added to the input end of the mixer.

4.5.4.3 Cross-modulation interference

When the input ends of the mixer simultaneously act on useful signal u_s, local oscillator signal u_L and interference signal u_M, the cross-modulation interference will occur due to the nonlinearity of mixing devices.

Assume that the U–I characteristics of mixing devices developed at a static operating point are as follows:

$$i = f(u) = a_0 + a_1 u + a_2 u^2 + a_3 u^3 + a_4 u^4 + \cdots$$

where

$$u = u_L + u_s + u_M = U_{Lm} \cos \omega_L t + U_{sm} \cos \omega_c t + U_{Mm} \cos \omega_M t$$

It is shown that the quadratic term ($2a_1 u_L u_s$ in the expansion), the fourth power term ($4a_4 u_L^3 u_s + 4a_4 u_L u_s^3 + 12a_4 u_L u_s u_M^2$ in the expansion) and the higher even-power term will produce the IF current component, where the amplitude of IF current generated by $12a_4 u_L u_s u_M^2$ is $3a_4 U_{Lm} U_{sm} U_{Mm}^2$, which is related to U_{Mm}. This indicates that the amplitude of current contains the envelope change of interference signal. In other words, this kind of interference is a kind of nonlinear distortion, which transports the envelope of the interference signal to the output IF signal, so it is called cross-modulation interference.

When this distortion exists, people not only hear the sound of the useful signal, but also hear the sound of the interference signal. However, when the useful signal station stops transmitting, the sound of interference signal disappears.

The methods of suppressing cross-modulation interference are to improve the selectivity of circuits at the front of a mixer. Second, square-law devices such as FETs can be selected, or make the transistor work as close as possible to the square-law region by properly choosing the transistor operating point current.

4.5.4.4 Intermodulation interference

When two interference signals u_{M1} and u_{M2} are acting on the input of mixer at the same time, the mixer may also produce intermodulation interference. If

$$u = u_L + u_s + u_{M1} + u_{M2} = U_{Lm} \cos \omega_L t + U_{sm} \cos \omega_c t + U_{M1m} \cos \omega_{M1} t + U_{M2m} \cos \omega_{M2} t$$

then the current i will contain the following combined frequency components:

$$f_{p,q,r,s} = |\pm p f_L \pm q f_c \pm r f_{M1} \pm s f_{M2}|$$

In addition to the useful IF component of $f_L - f_c = f_I$ ($p = q = 1, r = s = 0$), the following parasitic IF components may also exist on certain values of r and s that cause the output IF signal distortion of mixer, which is called intermodulation interference:

$$|\pm f_L \pm r f_{M1} \pm s f_{M2}| = f_I \tag{4.69}$$

It is obvious that the smaller the values of U_{M1m} and U_{M2m} are, the larger the amplitude of parasitic IF current component is, and the more serious the intermodulation interference is. If the frequencies f_{M1} and f_{M2} of two interference signals are very close to the frequency of the useful signal, the combined frequency component of two interference signals ($r = 1$, $s = 2$ or $r = 2$, $s = 1$) is likely to approach f_I when r and s are small:

$$f_L - (2f_{m1} - f_{m2}) \approx f_I \ or \ f_L - (2f_{m2} - f_{m1}) \approx f_I$$

That is

$$2f_{M1} - f_{M2} \approx f_c \ or \ 2f_{M2} - f_{M1} \approx f_c \qquad (4.70)$$

Therefore, this kind of intermodulation interference is the most serious. Because $r + s = 3$, this interference is called the third-order intermodulation interference, which is generated by $12a_4u_Lu_{M1}{}^2u_{M2}$ or $12a_4u_Lu_s\,u_{M1}u_{M2}{}^2$ of the fourth power term of u.

The methods of suppressing intermodulation interference are the same as those of suppressing cross-modulation interference. First, try to improve the selectivity of circuits at the front of mixer. Second, square-law devices such as FETs can be selected, or make the transistor work as close as possible to the square-law region by properly choosing the transistor operating point current.

Problems

4.1 Known that the power of carrier wave is 1,000 W, please find the whole power at $m_a = 1$ and $m_a = 0.7$, and the power of two sideband frequencies.

4.2 Given the expressions of amplitude-modulated wave, please draw their waveforms and spectrums.
(1) $(1 + \cos\Omega t)\cos\omega_c t$;
(2) $(1 + 0.5\cos\Omega t)\cos\omega_c t$;
(3) $\cos\Omega t\cos\omega_c t$ ($\omega_c = 5 \ \Omega$).

4.3 An expression of amplitude-modulated wave is

$$u = 25(1 + 0.7\cos2\pi \times 5{,}000 \ t - 0.3\cos22\pi \times 10^4 \ t)\sin2\pi \times 10^6 \ t)$$

try to find out the frequency and amplitude of components it contains.

4.4 Consider when the base modulation power amplifier has a maximum power, $I_{cmax} = 500$ mA, $2\theta = 120°$, $E_c = 12$ V, calculate P_S, P_{omax}, P_{Cmax} and η_C.

4.5 Given two signal voltages u_1 and u_2, the corresponding frequency parts are:

$u_1 = 2\cos2{,}000\pi t + 0.3\cos1{,}800\pi t + 0.3\cos2{,}200\pi t$(V)
$u_2 = 0.3\cos1{,}800\pi t + 0.3\cos2{,}200\pi t$(V)

Please answer the following questions:

(1) Are u_1 and u_2 modulated waves? Please list their mathematical expressions.

(2) Find out the power of sideband frequency and the whole power with unit resistor.

(3) Find out the frequency bandwidth B of the modulated wave.

4.6 The circuit of large signal peak envelope detector is shown in Ex.Fig. 4.1 with $R_L = 10\ \mathrm{k\Omega}$ and $m_a = 0.3$.

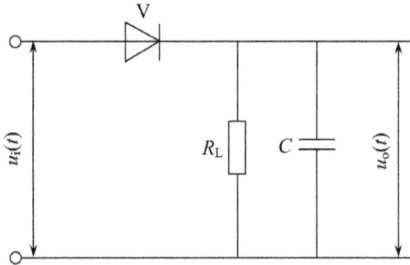

Ex.Fig. 4.1

(1) The frequency of carrier wave is 465 kHz and the highest frequency of a modulation signal is $F = 340$ Hz. How to select the capacitance of C? What is the approximate value of input impedance?

(2) Given $f_c = 30$ MHz and $F = 0.3$ MHz, how to select the capacitance of C? What is the approximate value of input impedance?

4.7 The demodulation circuit is shown in Ex.Fig. 4.2 with

$$u_i(t) = 5\cos 2\pi \times 465 \times 10^3\, t + 4\cos 2\pi \times 10^3\, t \cos 2\pi \times 465 \times 10^3\, t$$

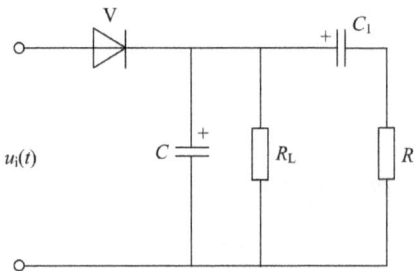

Ex.Fig. 4.2

The internal resistance of diode $r_D = 100\ \Omega$, $C = 0.01\ \mu F$ and $C_1 = 47\mu F$. In order to ensure no waveform distortion of output signal, please find

(1) The maximum load resistance R_L of demodulator

(2) The minimum input resistance R_i

4.8 The diode multiplier circuit is shown in Ex.Fig. 4.3. Assume that the I/V curve characteristics of diodes VD1 and VD2 are exactly the same, both with slope g_D and intercept 0 (pass the origin), $u_1(t) = U_{1m}\cos\omega_C t$, $u_2(t) = U_{2m}\cos\omega_L t$, $U_{2m} \gg U_{1m}$, $\omega_I = \omega_L - \omega_C$. Ignoring the load effect, please answer.

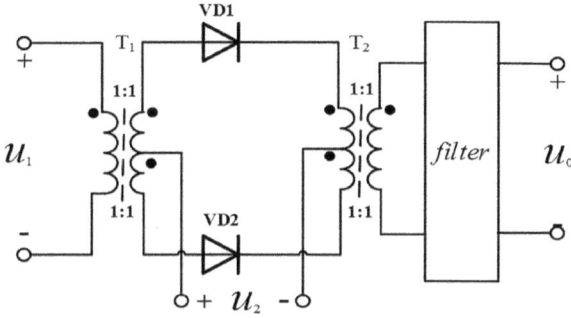

Ex.Fig. 4.3

(1) What type of filter should be used at u_o terminal?
(2) If the equivalent resistance of filter at the center frequency f_I is R_L, write the expression of output voltage u_o (t) and list the frequency components:
(3) Is the circuit able to realize frequency mixing? Why?

4.9 In a superheterodyne radio, if IF $f_I = f_L - f_c = 465$ kHz, try to analyze what kind of interference does the following phenomenon belong to. And how it was formed?
(1) When listening to the radio broadcast with a frequency of $f_c = 931$ kHz, hear a whistle about 1 kHz.
(2) When listening to the radio station with a frequency of $f_c = 550$ kHz, hear a strong radio station with a frequency of 1,480 kHz.
(3) When listening to the radio station with a frequency of $f_c = 1,480$ kHz, hear a strong radio station with a frequency of 740 kHz.

4.10 The operation frequency range of a heterodyne radio is from 535 to 1,605 kHz with the IF of 465 kHz. When listening to the radio station at $f_c = 700$ kHz, at which other frequencies may we receive the same radio signal (besides f_c)? Please list the two strongest frequencies and explain why that happens.

Chapter 5
Angle modulation and demodulation

5.1 Introduction

Frequency modulation (FM) and phase modulation (PM) could be classified as angle modulation. The frequency or phase of the carrier wave is modulated according to the modulation signal that people want to transmit. That is to say, the information of the modulation signal is hidden in the frequency or phase of the carrier wave. Once the modulated wave is received by the receiver, it can be demodulated. Then the modulation signal can be extracted.

For a high-frequency sinusoidal waveform, its mathematical expression should be as follows:

$$u_c(t) = U_{cm} \cos(\omega_c t + \varphi_0) \tag{5.1}$$

where U_{cm} is the amplitude of carrier wave, $\omega(t)$ is the angular frequency and φ_0 is the initial phase, $\omega t + \varphi_0$ represents the instantaneous phase:

$$\omega t + \varphi_0 = \varphi(t) \tag{5.2}$$

If this high-frequency expression is taken as the carrier wave, its basic parameters U_{cm}, $\omega_c(t)$ and $\varphi(t)$ can be tuned proportionally by a modulation signal. Normally the modulation signal wave has the following expression:

$$u_\Omega(t) = U_{\Omega m} \cos \Omega t \tag{5.3}$$

As introduced in the previous chapter, when U_{cm} was modulated by a modulation signal, the carrier wave will become an amplitude modulated (AM) wave with the following expression:

$$u_{AM}(t) = U_{cm}(1 + m_a \cos \Omega t) \cos(\omega_c t + \varphi_0) \tag{5.4}$$

To simplify the case, let $\varphi_0 = 0$. Therefore eq. (5.4) is rewritten as

$$u_{AM}(t) = U_{cm}(1 + m_a \cos \Omega t) \cos(\omega_c t) \tag{5.5}$$

Similarly, when $\omega(t)$ is modulated by $u_\Omega(t)$, frequency of the carrier wave varies linearly with the modulation signal as

$$\omega(t) = \omega_c + k_f u_\Omega(t) \tag{5.6}$$

The information of $u_\Omega(t)$ is embedded into frequency variations of the carrier wave. This is the so-called FM.

When $\varphi(t)$ is modulated by $u_\Omega(t)$, phase of the carrier wave varies linearly with the modulation signal as

https://doi.org/10.1515/9783110593822-005

$$\varphi(t) = \omega_c t + k_p u_\Omega(t) \tag{5.7}$$

The information of $u_\Omega(t)$ is embedded into phase variations of the carrier wave. This is the so-called PM.

Different from AM, which is a process of linear frequency shifting, both FM and PM lead to nonlinear frequency conversions. In frequency spectrum, the FM or PM signals cover much wider bandwidth than AM (Fig. 5.1). Moreover, the modulated waveforms are different from AM, FM and PM. As we know, energy or power of a wave is closely related to its voltage amplitude. From the view of power consumption, FM or PM is just the power redistribution of the previous carrier wave, without changing the total power after modulation (i.e., amplitude is unchanged). This allows FM or PM to be much better in antiinterference properties. FM has been widely used in FM broadcast, TV and communication system. PM is used mainly in digital communication systems.

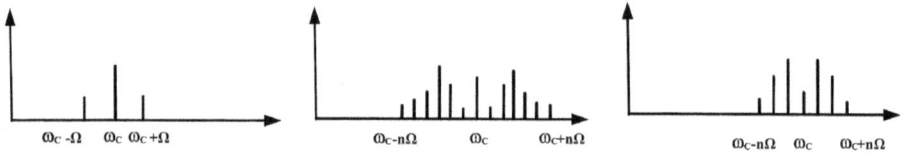

Fig. 5.1: Spectrum of the AM, FM and PM.

For FM, the demodulation process is also called frequency discrimination. For PM, that is called phase detection.

5.2 Properties of angle modulation wave

5.2.1 Instantaneous frequency and instantaneous phase

For a sinusoidal wave with the following expression $u_c(t) = U_{cm} \cos(\omega t + \varphi_0)$, the instantaneous frequency $\omega(t)$ and the instantaneous phase $\varphi(t)$ are related by

$$\varphi(t) = \int_0^t \omega(t)dt + \varphi_0 \tag{5.8}$$

Phase is a dimensionless physical quantity. $\int_0^t \omega(t)dt$ represents the total angle or phase that a vector travelled from $t = 0$ to time t when the vector is scanning with an angular frequency of $\omega(t)$.

By taking derivative at both sides of eq. (5.8), we get the relationship between w and φ in the differential form of

$$w(t) = \frac{d\varphi(t)}{dt} \tag{5.9}$$

The instantaneous frequency $w(t)$ is the changing rate of the instantaneous phase $\varphi(t)$.

5.2.2 Mathematical expression of FM and PM

5.2.2.1 FM
For simplicity we start by assuming that the modulation signal is a monofrequency with the expression of $u_\Omega(t) = U_{\Omega m} \cos \Omega t$ (Ω is the modulation signal's frequency). As for the carrier wave, the expression should be $u_c(t) = U_{cm} \cos w_c t$. As explained earlier, FM wave indicates that the instantaneous frequency of the carrier wave should change linearly with the modulation signal, which can be expressed as $w(t) = w_c + k_f u_\Omega(t) = w_c + \Delta w(t)$ (k_f is a FM coefficient with the dimension of rad/(V·s)). k_f describes the frequency offset caused by unit amplitude of the modulation signal. $k_f u_\Omega(t) = \Delta w(t)$ is the frequency offset deviating from its central carrier wave frequency w_c at the moment of t. When $\cos \Omega t = 1$, $\Delta w(t)$ reaches its maximum value of $\Delta w_f = k_f U_{\Omega m}$.

According to eq. (5.8), the instantaneous phase of the FM wave is

$$\varphi(t) = \int_0^t w(t)dt = \int_0^t [w_c + k_f u_\Omega(t)]dt = w_c t + k_f \frac{U_{\Omega m}}{\Omega} \sin \Omega t \tag{5.10}$$

$$\Delta\varphi(t) = k_f \frac{U_{\Omega m}}{\Omega} \sin \Omega t \tag{5.11}$$

Equation (5.11) is the instantaneous phase shift. The maximum phase shift is defined as FM index m_f:

$$m_f = k_f \frac{U_{\Omega m}}{\Omega} \tag{5.12}$$

Then a generalized or standard expression for FM signal is obtained as

$$u_{FM}(t) = U_{cm} \cos\left(w_c t + k_f \frac{U_{\Omega m}}{\Omega} \sin \Omega t\right) = U_{cm} \cos(w_c t + m_f \sin \Omega t) \tag{5.13}$$

FM signal is with unchanged amplitude and time-varying phase (or frequency). Index m_f is usually larger than 1. Larger m_f usually denotes better antiinterference property.

5.2.2.2 PM

For PM, the instantaneous phase changes linearly with the modulation signal, hence the instantaneous phase of the PM wave is

$$\varphi(t) = w_c t + k_p u_\Omega(t) = w_c t + \Delta\varphi(t) \tag{5.14}$$

where $w_c t$ is the phase of carrier wave (assume $\varphi_0 = 0$) and k_p is a coefficient for PM also with the dimension of rad/(V·s). k_p describes the phase shift caused by unit amplitude of the modulation signal. For PM wave, the maximum phase shift m_p can be observed directly.

$$m_p = k_p U_{\Omega m} \tag{5.15}$$

m_p is also called the PM index. According to eq. (5.9), the instantaneous frequency of the PM wave is calculated by

$$w(t) = \frac{d\varphi(t)}{dt} = w_c + \Omega k_p U_{\Omega m} \sin(\Omega t + \pi) = w_c + \Delta w(t) \tag{5.16}$$

where $\Delta w(t)$ is called the instantaneous frequency shift. Its maximum value is $\Delta w_p = \Omega k_p U_{\Omega m}$.

Then a standard expression for a PM wave is given by

$$u_{PM}(t) = U_{cm} \cos(w_c t + k_p U_{\Omega m} \cos \Omega t) = U_{cm} \cos(w_c t + m_p \cos \Omega t) \tag{5.17}$$

5.2.2.3 Comparison between FM and PM

The comparison between FM and PM is shown in Table 5.1.

Table 5.1: Comparison of FM and PM.

	FM	PM
Instantaneous frequency	$w_c + k_f u_\Omega(t)$	$w_c + \Omega k_p U_{\Omega m} \sin(\Omega t + \pi)$
Instantaneous phase	$w_c t + k_f \frac{U_{\Omega m}}{\Omega} \sin \Omega t$	$w_c t + k_p u_\Omega(t)$
Maximum frequency shift	$\Delta w_f = k_f U_{\Omega m}$	$\Delta w_p = \Omega k_p U_{\Omega m}$
Maximum phase shift	$m_f = k_f \frac{U_{\Omega m}}{\Omega}$	$m_p = k_p U_{\Omega m}$
Mathematical expression	$u_{FM}(t) = U_{cm} \cos\left(w_c t + k_f \frac{U_{\Omega m}}{\Omega} \sin \Omega t\right)$ $= U_{cm} \cos(w_c t + m_f \sin \Omega t)$	$u_{PM}(t) = U_{cm} \cos(w_c t + k_p U_{\Omega m} \cos \Omega t)$ $= U_{cm} \cos(w_c t + m_p \cos \Omega t)$

We can draw some conclusions about FM and PM. $\Delta\omega$ and m of angle modulation signal versus Ω are shown as Fig. 5.2. m_x is the modulation index of FM and PM, either of which stands for the maximum phase shift. For PM, it is constant when the amplitude of the modulation signal is constant; while for FM, it is proportional to $1/\Omega$. For FM, the maximum frequency shift $\Delta\omega_f = k_f U_{\Omega m}$ is a constant. That means the bandwidth of FM is not related to Ω, the frequency of the modulation signal. For PM, the maximum frequency shift $\Delta\omega_p = \Omega k_p U_{\Omega m}$ is proportional to Ω. That means its bandwidth is proportional to the frequency of the modulation signal. For FM and PM, $\Delta\omega_x = m_x\Omega$. For the same modulation signal and carrier wave, FM waveform resembles that of PM largely. The main difference is that phase of FM wave is delayed by $\pi/2$ compared to PM wave.

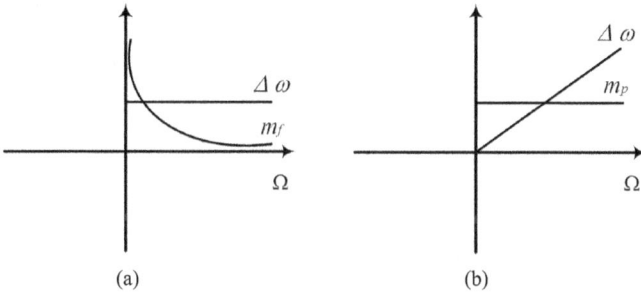

Fig. 5.2: Functions of $\Delta\omega$ and M versus Ω of FM and PM, (a) FM, (b) PM.

5.2.3 Frequency spectrum and bandwidth of FM and PM

The expression of an FM signal is given as $u_{FM}(t) = U_{cm}\cos(\omega_c t + m_f \sin\Omega t)$. According to the mathematical rule of sum to product, $\cos(\alpha+\beta) = \cos\alpha\cos\beta - \sin\alpha\sin\beta$.

Therefore, $u_{FM}(t) = U_{cm}[\cos(\omega_c t)\cos(m_f \sin\Omega t) - \sin(\omega_c t)\sin(m_f \sin\Omega t)]$. In the expression, $\cos(m_f \sin\Omega t)$ and $\sin(m_f \sin\Omega t)$ could be expanded into a series of sums of Bessel functions of the first kind, $J_n(m_f)$. The image and the listed coefficients are shown in Fig. 5.3 and Table 5.2, respectively,

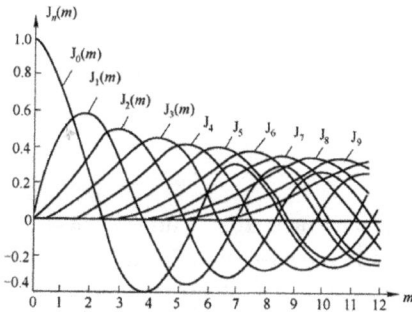

Fig. 5.3: Bessel functions of the first kind.

Table 5.2: Listed $J_n(m_f)$ coefficients.

m_f	$J_0(m_f)$	$J_1(m_f)$	$J_2(m_f)$	$J_3(m_f)$	$J_4(m_f)$	$J_5(m_f)$	$J_6(m_f)$	$J_7(m_f)$	$J_8(m_f)$	$J_9(m_f)$
0.01	1.00	0.005								
0.20	0.99	0.100								
0.50	0.94	0.24	0.03							
1.00	0.72	0.44	0.11	0.02						
2.00	0.22	0.58	0.35	0.13	0.03					
3.00	0.26	0.34	0.49	0.31	0.13	0.04	0.01			
4.00	0.39	0.06	0.36	0.48	0.28	0.13	0.05	0.01		
5.00	0.18	0.33	0.05	0.36	0.39	0.26	0.13	0.05	0.02	
6.00	0.15	0.28	0.24	0.11	0.36	0.36	0.25	0.13	0.06	0.02

$$\cos(m_f \sin \Omega t) = J_0(m_f) + 2J_2(m_f)\cos 2\Omega t + 2J_4(m_f)\cos 4\Omega t + \cdots$$

$$= J_0(m_f) + 2\sum_{n=1}^{\infty} J_{2n}(m_f)\cos 2n\Omega t$$

$$\sin(m_f \sin \Omega t) = 2J_1(m_f)\sin \Omega t + 2J_3(m_f)\sin 3\Omega t + 2J_5(m_f)\sin 5\Omega t + \cdots$$

$$= 2\sum_{n=1}^{\infty} J_{2n-1}(m_f)\sin(2n-1)\Omega t$$

$$\therefore u_{FM}(t) = U_{cm}\left[\cos(\omega_c t)J_0(m_f) + 2\cos(\omega_c t)\sum_{n=1}^{\infty} J_{2n}(m_f)\cos 2n\Omega t\right.$$

$$\left. - 2\sin(\omega_c t)\sum_{n=1}^{\infty} J_{2n-1}(m_f)\sin(2n-1)\Omega t\right]$$

$$= U_{cm}[J_0(m_f)\cos(\omega_c t) + J_2(m_f)\cos(\omega_c + 2\Omega)t + J_2(m_f)\cos(\omega_c - 2\Omega)t + \cdots$$

$$+ J_1(m_f)\cos(\omega_c + \Omega)t - J_1(m_f)\cos(\omega_c - \Omega)t + \cdots]$$

$$= U_{cm}\sum_{n=-\infty}^{\infty} J_n(m_f)\cos(\omega_c + n\Omega)t \qquad (5.18)$$

The last step used the following product to sum rules as well as the property of Bessel functions of the first kind.

$$\begin{cases} \cos\alpha\cos\beta = \dfrac{1}{2}\cos(\alpha+\beta) + \dfrac{1}{2}\cos(\alpha-\beta) \\ \sin\alpha\sin\beta = \dfrac{1}{2}\cos(\alpha-\beta) - \dfrac{1}{2}\cos(\alpha+\beta) \end{cases}$$

$$\begin{cases} J_{-n}(m_f) = J_n(m_f), & n = 0, 2, 4, \ldots \\ J_{-n}(m_f) = -J_n(m_f), & n = 1, 3, 5, \ldots \end{cases}$$

From eq. (5.18) we can get some clue about the frequency spectrum of an FM wave [17].
(1) There are innumerable frequency components in the spectrum. The amplitude of each frequency component is proportional to $J_n(m_f)$. $J_n(m_f)$ tends to be smaller as

the order n gets larger. Therefore, in the spectrum, the frequencies further apart from central ω_c are usually lower in amplitude.

(2) The power of a FM signal P_{FM} includes the power of the carrier wave power P_{ca} and of all side bands P_{sb}, $P_{FM} = P_{ca} + P_{sb}$. At the same time, power conserves before and after FM process. FM technique is essentially a power redistribution process. Furthermore, it is regularly distributed carrier power to sideband spectral components. This can be supported by the normalization property of the Bessel function:

$$\sum_{n=-\infty}^{\infty} J_n^2 (m_f) = 1$$

By following the same mathematical procedure, a PM signal can be written as

$$u_{PM}(t) = U_{cm}[J_0(m_p)\cos(\omega_c t) - J_2(m_p)\cos(\omega_c - 2\Omega)t + J_2(m_p)\cos(\omega_c + 2\Omega)t$$
$$+ J_4(m_p)\cos(\omega_c - 4\Omega)t + J_4(m_p)\cos(\omega_c + 4\Omega)t - \cdots] \tag{5.19}$$

The above equation is deduced according to the following expressions:

$$\cos(m_p \cos \Omega t) = J_0(m_p) - 2J_2(m_p)\cos 2\Omega t + 2J_4(m_p)\cos 4\Omega t + \cdots$$
$$= J_0 (m_p) + 2\sum_{n=1}^{\infty} (-1)^n J_{2n}(m_p)\cos 2n\Omega t$$

$$\sin(m_p \cos \Omega t) = -2J_1(m_p)\cos \Omega t + 2J_3(m_p)\cos 3\Omega t - 2J_5(m_p)\cos 5\Omega t + \cdots$$
$$= -2\sum_{n=1}^{\infty} (-1)^n J_{2n-1}(m_p)\cos(2n-1)\Omega t$$

According to expressions, the spectrum of PM covers a wide range of $\omega_c \pm n\Omega$ in which n can be any positive integer.

The spectrum of the angle modulation signal theoretically includes infinite frequency components, and the amplitude of the nth component is proportional to $J_n(m)$. The magnitude of $J_n(m)$ decreases quickly as n gets larger. Observed from Table 5.2, when $n > m + 1$, the magnitude of $J_n(m)$ is lower than 0.1. In engineering practices, frequency components whose amplitude is lower than 10% of the carrier waveform could be neglected.

Consequently, the effective bandwidth of an FM signal is approximated as

$$B_f = 2nF \approx 2(m_f + 1)F \tag{5.20}$$

For $m_f = \dfrac{k_f}{\Omega} U_{cm}$, the bandwidth can be further written as

$$B_f \approx 2(m_f + 1)F = 2\Delta f + 2F \tag{5.21}$$

In reality the modulation signal is not a monochromatic or monofrequency signal but covers a certain bandwidth from F_{min} to F_{max}. The corresponding FM signal's effective bandwidth is $B_f = 2nF_{max} \approx 2(m_f + 1)F_{max}$. This equation is applicable to m_f which is larger than 1. For cases of which m_f is much less than 1, $B_f \approx 2F$.

The effective bandwidth of PM signals is obtained with the similar form of $B_p \approx 2(m_p + 1)F_{max}$. When the frequency Ω of the modulation signal changes, the PM index m_p changes proportionally. As a result, the bandwidth becomes wider when Ω increases.

5.2.4 Power distribution of the angle modulation wave

The power of FM or PM signals is calculated as the total power of the carrier wave and the modulation signal. The amplitude of either FM or PM signals stays the same with the original carrier wave, indicating the total power remains unchanged before and after angle modulation. The existence of sideband components does not require any extra power input from the circuit but acquires the power from the carrier wave itself. To put it in another way, angle modulation is a power redistribution process between the carrier wave and the emerging sideband frequencies [18].

5.3 Realization of FM

5.3.1 Method of the frequency modulation

FM can be realized by two different approaches, direct FM and indirect FM. The essence of FM is to control the carrier frequency regularly by a modulation signal.

For direct FM, the modulation signal is sent into an oscillator, which consists of variable reactance device, such as voltage-controlled varactors. When the voltage (i.e., amplitude) of the modulation signal varies the capacitance changes accordingly, resulting in the oscillation frequency being tuned in a controlled way. A circuit with direct FM function is completed.

For indirect FM, the oscillator's frequency is fixed. But the phase of the carrier wave is modulated by the modulation signal. A phase-to-frequency translation network is used afterwards to realize FM in an indirect way.

5.3.2 Performance specifications of the FM circuit

(1) Modulation characteristics. For a modulated signal, the ratio between frequency offset Δf and the center frequency (or carrier wave frequency) f_c is defined as the modulation characteristics. This ratio is a function of modulation signal u_Ω.

$$\frac{\Delta f}{f_c} = f(u_\Omega) \tag{5.22}$$

Where u_Ω is the modulation signal voltage. Δf should vary with u_Ω linearly in the ideal case. However, nonlinear distortion is almost unavoidable.

(2) Modulation sensitivity S. Sensitivity S is defined as the frequency shift caused by unit modulation voltage.

$$\frac{\Delta f}{\Delta u} = S \qquad (5.23)$$

(3) The maximum frequency shift Δf_m. Δf_m is the maximum possible frequency shift or offset value for a modulated signal.

(4) Carrier wave frequency stability. It is defined as the ratio between the carrier wave's frequency shift Δf and original carrier wave frequency f_c during an certain interested time period.

5.4 FM circuit

5.4.1 FM based on varactors

5.4.1.1 Varactors

A varactor is a voltage-controlled component with variable reactance. Its junction capacitance changes remarkably with a reverse bias voltage. The images, circuit symbol and equivalent circuit model are shown in Fig. 5.4.

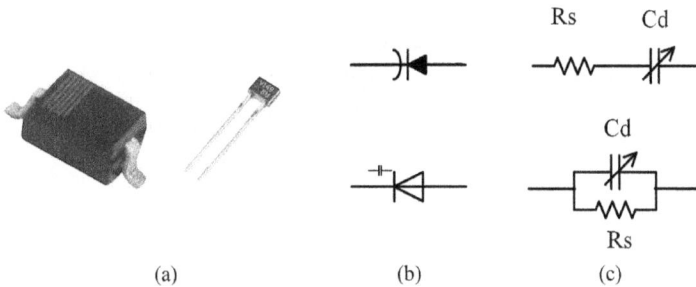

Fig. 5.4: Varactor and equivalent circuit, (a) Images of varactors, (b) circuit symbol, (c) the equivalent circuit model.

A measured result of how junction capacitance C_d changes with the reversely applied voltage V_R is displayed (Fig. 5.5).

The C–V relationships of varactors can be summarized mathematically as

$$C_j = \frac{C_0}{\left(1 + \frac{u_R}{U_D}\right)^n} \qquad (5.24)$$

Where c_0 is the capacitance when no bias voltage is applied, which is determined by the physical properties of PN junction. U_D is the barrier voltage of the PN junction. u_R

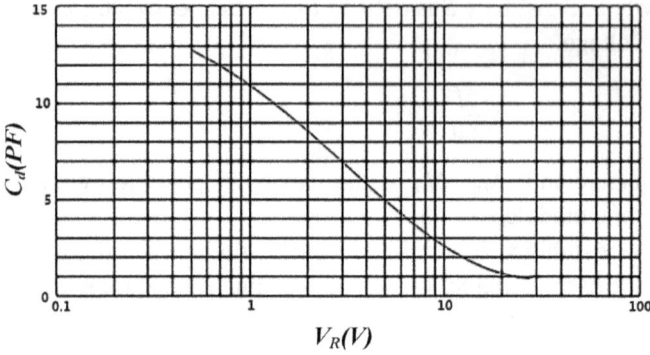

Fig. 5.5: *C–V* characteristics of a varactor (Philips BB181 @25 °C).

is the absolute value of the bias voltage. n is a power index depending on the type of PN junction. Typical values are that $n = 1/3$ for graded junctions and $n = 1/2$ for abrupt junctions (also known as step junctions).

5.4.1.2 Principle of varactor-based FM circuit

Figure 5.6 is the example FM circuit realized by a varactor. In Fig. 5.6(a), C_3 is a high-frequency coupling capacitor which can be ignored in frequency calculation due to its large capacitance. L_D is an RF choke, stopping the high-frequency signals from disturbing the power supply E_C while allowing E_C and the low FM signal passing through. The oscillation frequency of the circuit is $\omega = \frac{1}{\sqrt{LC_\Sigma}}$ with $C_\Sigma = C_j + \frac{C_1 C_2}{C_1 + C_2}$. To make it simpler, we can let $C_\Sigma \approx C_j$ and get $\omega = \frac{1}{\sqrt{LC_j}}$. R_1 and R_2 are

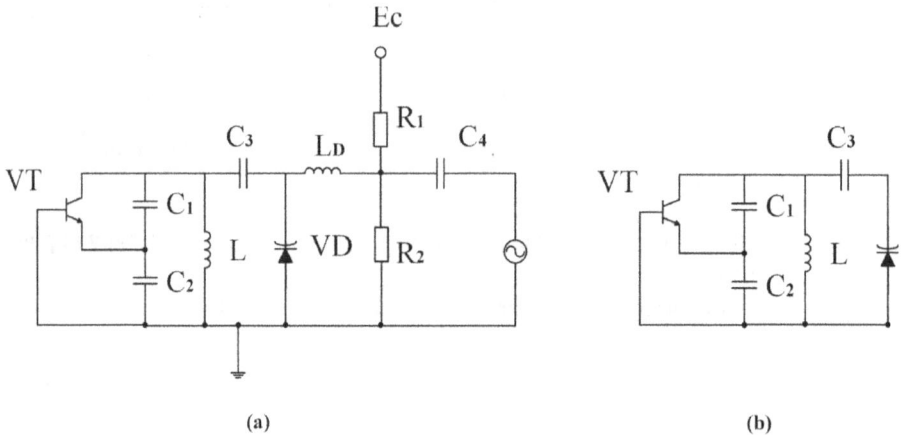

(a)   (b)

Fig. 5.6: (a) A typical FM circuit, (b) equivalent AC circuit.

used to set the quiescent DC bias for the varactor: $U_Q = E_c \cdot \frac{R_2}{R_1 + R_2}$. The total voltage applied to the varactor is $u(t) = U_Q + u_\Omega(t)$.

The capacitance of the varactor C_j is

$$C_j = \frac{C_0}{\left[1 + \frac{U_Q + u_\Omega(t)}{U_D}\right]^n} = \frac{C_0}{\left(1 + \frac{U_Q + u_{\Omega m}\cos\Omega t}{U_D}\right)^n} = C_0\left(1 + \frac{U_Q + u_m\cos\Omega t}{U_D}\right)^{-n}$$

$$= C_0\left(\frac{U_Q + U_D}{U_D} + \frac{u_{\Omega m}\cos\Omega t}{U_D}\right)^{-n}$$

$$= C_0\left[\left(\frac{U_Q + U_D}{U_D}\right)\left(1 + \frac{u\Omega_{\Omega m}}{U_Q + U_D}\cos\Omega t\right)\right]^{-n}$$

$$= \frac{C_0}{\left(1 + \frac{U_Q}{U_D}\right)^n}\left(1 + \frac{u_{\Omega m}}{U_Q + U_D}\cos\Omega t\right)^{-n} \tag{5.25}$$

where $\dfrac{C_0}{\left(1 + \frac{U_Q}{U_D}\right)^n}$ is recognized for the static junction capacitance C_{j0}. Let $m = \dfrac{u_{\Omega m}}{U_Q + U_D}$

$$C_j = C_{j0}(1 + m\cos\Omega t)^{-n} \tag{5.26}$$

Substitute the expression of C_j into the expression of ω

$$\omega = \frac{1}{\sqrt{LC_j}} = \frac{1}{\sqrt{LC_{j0}(1 + m\cos\Omega t)^{-n}}} \tag{5.27}$$

$$= \omega_0(1 + m\cos\Omega t)^{\frac{n}{2}}$$

(1) If $n = 2$,

$$\omega = \omega_0(1 + m\cos\Omega t) = \omega_0 + \omega_0\frac{u_{\Omega m}}{U_0 + U_D}\cos\Omega t = \omega_0 + k_f\cos\Omega t \tag{5.28}$$

where $k_f = \omega_0 \cdot \frac{u_{\Omega m}}{U_Q + U_D}$. Apparently, it is linear FM in which the modulated frequency ω is linearly dependent on the modulation signal u_Ω.

(2) If $n \neq 2$, eq. (5.27) could be broken down by the Taylor expansion, $(1 + x)^n = 1 + nx + (1 + x)^n = 1 + nx + \frac{n(n-1)}{2!}x^2 + \cdots$. Then $\omega = \omega_0(1 + m\cos\Omega t)^{\frac{n}{2}}$ is expanded as

$$\omega = \omega_0(1 + m\cos\Omega t)^{\frac{n}{2}} = \omega_0\left[1 + \frac{nm}{2}\cos\Omega t + \frac{\frac{n}{2}\left(\frac{n}{2} - 1\right)}{2!}(m\cos\Omega t)^2 + \cdots\right] \tag{5.29}$$

Neglecting higher order terms, we obtain the modulated frequency to be

$$\omega = \omega_0\left[1 + \frac{n}{8}\left(\frac{n}{2} - 1\right)m^2\right] + \frac{n}{2}m\omega_0\cos\Omega t + \frac{n}{8}\left(\frac{n}{2} - 1\right)\omega_0 m^2\cos 2\Omega t$$

$$= (\omega_0 + \Delta\omega_0) + \Delta\omega_m\cos\Omega t + \Delta\omega_{2m}\cos 2\Omega t \tag{5.30}$$

Where $\Delta\omega_0 = \frac{n}{8}\left(\frac{n}{2}-1\right)m^2\omega_0$, $\Delta\omega_m = \frac{n}{2}m\omega_0$ and $\Delta\omega_{2m} = \frac{n}{8}\left(\frac{n}{2}-1\right)m^2\omega_0$.

From eq. (5.30), several conclusions can be drawn.

① For $n \neq 2$ cases, frequency ω is no longer linear with u_Ω. The central frequency is not ω_0 but $(\omega_0 + \Delta\omega_0)$ which is shifted by the amount of $\Delta\omega_0 = \frac{n}{8}\left(\frac{n}{2}-1\right)m^2\omega_0$.

② The maximum frequency shift is $\Delta\omega_m = \frac{n}{2}m\omega_0$, which is proportional to n, m and ω_0.

③ Higher frequencies such as 2Ω term in eq. (5.30) emerge. If higher order terms are too prominent to be ignored due to nonlinearity, nonlinear distortion effect becomes obvious.

5.4.1.3 Practical circuit of varactor-based FM

In Fig. 5.7 we present a Seiler oscillator built by transistor VT to realize FM. The modulation signal is introduced into the circuit through C_7, and the modulated signal or output is obtained from C_2. A pair of varactors VD_1 and VD_2 are connected back to back to reduce the fluctuation on the serial capacitance caused by high-frequency signals. If the bias voltage for one varactor suddenly changes, then the bias condition for the other varactor varies in the opposite way keeping the equivalent capacitance roughly unchanged.

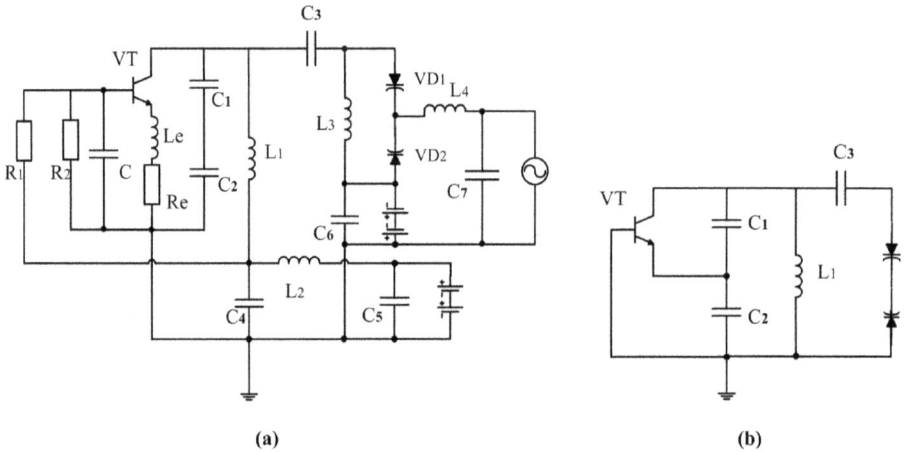

Fig. 5.7: The FM with varactor, (a) practical circuit, (b) AC equivalent circuit.

5.4.2 FM circuit based on VCO

The block diagram of FM based on voltage-controlled oscillator (VCO) is shown in Fig. 5.8. The output frequency of the VCO changes regularly with the control

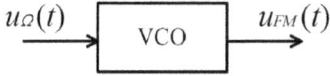

Fig. 5.8: Block diagram of FM based on VCO.

voltage as $w(t) = w_c + k_f u_\Omega(t)$. As the modulation signal is controlling VCO, the output signal should be

$$u_{FM}(t) = U_{cm} \cos\left[w_c t + \int k_f u_\Omega(t)dt\right] = U_{cm} \cos(w_c t + m_f \sin \Omega t) \tag{5.31}$$

Then FM is realized in principle.

5.4.3 FM circuit based on quartz crystal resonator

5.4.3.1 Basic principle
A typical Pierce oscillator is shown in Fig. 5.9 with the oscillating frequency of

$$f_0 = f_q\left[1 + \frac{C_g}{2(C_L + C_0)}\right] \tag{5.32}$$

where f_q is the series resonance frequency of the crystal. C_g and c_0 are the dynamic and static capacitance of the crystal, respectively. C_L is the total capacitance of c_1, c_2 and c_j, which are in series, $C_L = \frac{1}{\frac{1}{C_1} + \frac{1}{C_2} + \frac{1}{C_j}}$. If c_0 is substituted by a varactor and controlled by the modulation signal, FM can be realized. Quartz crystal oscillator must work in the inductive region, so the circuit frequency must be between f_s and f_p. Its relative frequency shift (or deviation) range is $\frac{f_p - f_s}{f_0} = 10^{-3} - 10^{-4}$ which is quite limited for many application purposes. To expand the frequency shift range there are at least two ways. We can either connect a small capacitance in series or use Π-shape network to expand the inductive range of the quartz crystal [19].

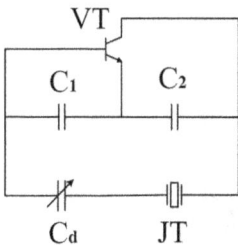

Fig. 5.9: Circuit model of FM based on a Pierce oscillator.

5.4.3.2 Practical circuit of quartz crystal FM
A practical circuit of FM with Pierce oscillator and its AC equivalent circuit are both shown in Fig. 5.10. In this circuit, JT is used as a large inductor. JT, VD, C_1 and C_2

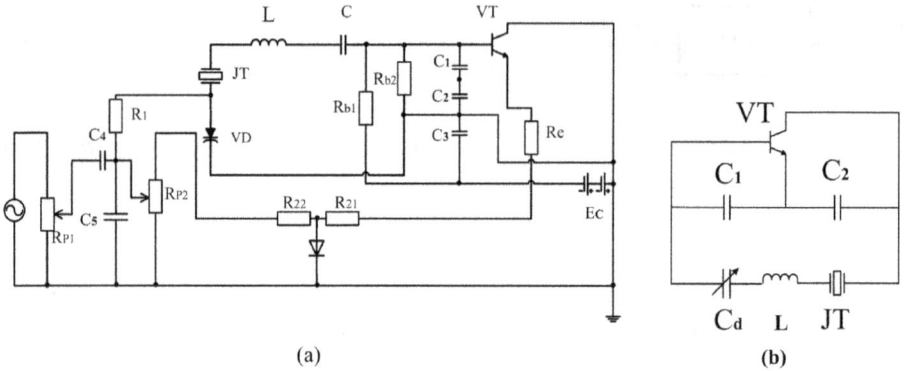

Fig. 5.10: Pierce oscillator, (a) practical FM circuit, (b) AC equivalent circuit.

form the oscillator loop together. When the varactor capacitance changes the series resonant frequency, the crystal will move as well and change the overall oscillation frequency. FM is realized this way. With high-frequency stability of quartz crystal, this whole FM system also benefits from fairly high-frequency stability.

5.4.4 Indirect FM circuit

5.4.4.1 Principle of indirect FM

Indirect FM can be realized by PM first and then signal conversion. The process of indirect FM is shown in Fig. 5.11. In the above block diagram a stable crystal oscillator is used as the main oscillator to generate the carrier wave. Then three steps are required in order to obtain the indirect FM signal.

(1) The modulation signal $u_\Omega(t)$ needs to be integrated to get $\int u_\Omega(t)dt$.

(2) $\int u_\Omega(t)dt$ is used as the phase control signal to perform PM to the carrier wave. After that a narrow FM wave $u_{FM}(t)$ is generated.

(3) Multiple class of frequency doubling and mixing is performed to realize wideband FM with proper and satisfying center frequency and frequency shift.

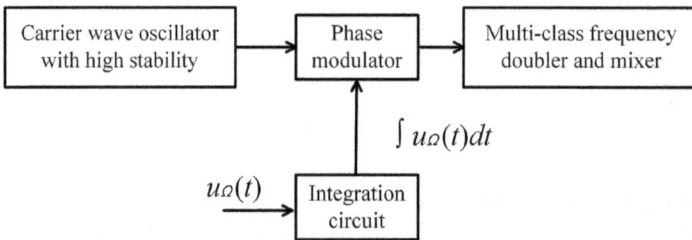

Fig. 5.11: Block diagram of indirect FM principle.

5.4.4.2 PM circuit

Varactor PM circuits are widely used. For a parallel LC resonance loop, the loop impedance is

$$Z(j\omega) = \frac{R_e}{\sqrt{1 + jQ_L \frac{2(\omega - \omega_0)}{\omega_0}}} = Z(\omega)e^{j\varphi_z(\omega)}$$

where R_e is the resonance impedance.

$$Z(w) = \frac{R_e}{\sqrt{\left\{1 + \left[Q_L \frac{2(\omega - \omega_0)}{\omega_0}\right]^2\right\}}} \tag{5.33}$$

$$\varphi_z(\omega) = -\arctan\left[Q_L \frac{2(\omega - \omega_0)}{\omega_0}\right], \quad Q_L = \frac{R_e}{\omega_0 L}$$

$$\omega(t) = \frac{1}{\sqrt{LC_j}} = \omega_0(1 + m\cos\Omega t)^{\frac{n}{2}} \approx \omega_0\left(1 + \frac{n}{2}m\cos\Omega t\right)$$

$$\therefore \Delta\omega(t) \approx \frac{n}{2}\omega_0 m \cos\Omega t \tag{5.34}$$

According to the characteristic curves shown in Fig. 5.12, when $\Delta\varphi < \frac{\pi}{6}$ $\tan\Delta\varphi \approx \Delta\varphi$,

$$\therefore \Delta\varphi = -Q_L\frac{2(\omega - \omega_0)}{\omega_0} = -Q_L\frac{2\Delta\omega}{\omega_0} \tag{5.35}$$

Substituting eq. (5.34) into eq. (5.35) we get

$$\Delta\varphi = -Q_L nm\cos\Omega t = -Q_L n\frac{u_m}{U_Q + U_D}\cos\Omega t \tag{5.36}$$

In conclusion phase shift is approximately linear with the modulation signal, which indicates that the circuit is a PM circuit. Suppose the modulation signal was integrated already then what we finally get is actually an FM signal by comparing PM and FM definitions.

In Fig. 5.13, an actual indirect FM circuit is shown for illustration. A varactor C_j together with C_3 and L constitutes a resonant network (PM network) for phase shifting. Capacitance is chosen as $C_j \ll C_3$ so that the resonance frequency is dominated by C_j. In the circuit, C_1, C_2 and C_3 ($C_1 = C_2 = C_3 = 0.001\ \mu F$) are considered as short to the high-frequency signal and open to the modulation signal. R_1 and R_2 are the blocking resistors of input and output side. DC bias of C_j is determined by the power supply and the integrated modulation signal jointly. The integration operation is realized by R_3, R_4 and C_4, with the requirement of

$$\frac{1}{\Omega C} \gg R$$

Amplitude-frequncy characteristics

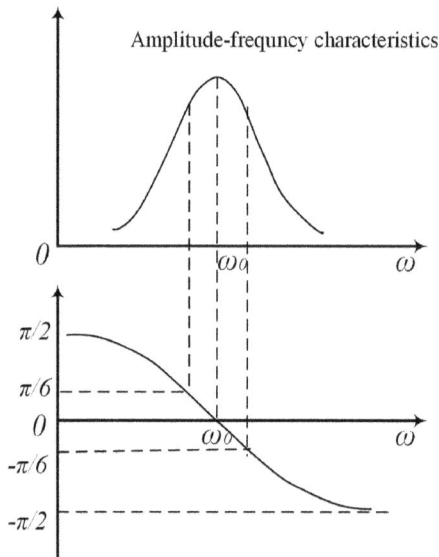

Fig. 5.12: Impedance amplitude and phase characteristics of LC parallel circuit.

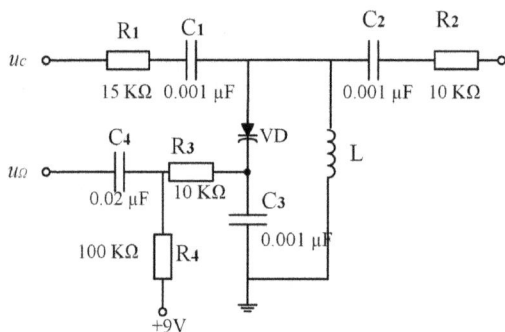

Fig. 5.13: An indirect FM circuit example.

At the output end

$$U'_{\Omega m} = U_{\Omega m} \frac{\frac{1}{\Omega C}}{\sqrt{R^2 + \left(\frac{1}{\Omega C}\right)^2}} \approx U_{\Omega m} \frac{\frac{1}{\Omega C}}{R} = U_{\Omega m} \frac{1}{RC} \frac{1}{\Omega} \propto \frac{U_{\Omega m}}{\Omega} \tag{5.37}$$

When $U'_{\Omega m}$ is used as the control signal followed by PM network, according to eq. (5.36)

$$\Delta \varphi = -Q_L n m \cos \Omega t = -Q_L n \frac{u'_{\Omega m}}{U_Q + U_D} \cos \Omega t \propto \frac{U_{\Omega m}}{\Omega} \tag{5.38}$$

It reveals the typical feature of FM where phase varies linearly with the control signal.

However, to ensure good linearity $\Delta\varphi$ should be kept smaller than $\pi/6$ in this PM circuit. That limits maximum frequency shift greatly for practical use. Consequently, frequency doubling technique is usually needed to enlarge modulation depth. PM is performed at lower frequencies, and carrier frequency is doubled for several times. Then maximum frequency shift can be expanded to the desired level with convenience. Another approach is to use multiple PM networks and get the sum of phase shift. Modulation depth is also enlarged this way.

Figure 5.14 is another indirect FM circuit example with single level of PM network and a varactor. The parallel resonant loop is made up by L, C_1, C_2 and C_j; C_3, C_4 and C_5 are coupling capacitors. The carrier wave signal is sent in through C_3. The modulation signal feeds the circuit through C_5. The PM signal output is obtained through C_4.

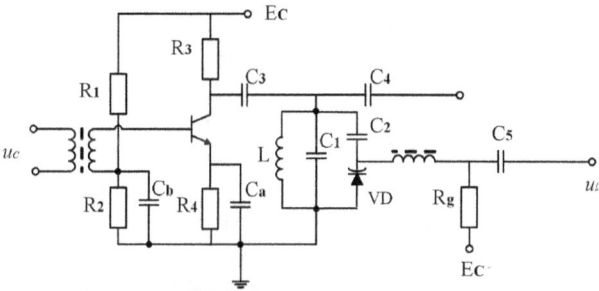

Fig. 5.14: An indirect FM circuit example with single level of PM network and a varactor.

In Fig. 5.15 another indirect FM circuit with multiple PM networks (three levels) is presented. Each of the three levels of PM network contains an independent varactor and is controlled by the same modulation signal. Q factor of each level is adjusted by

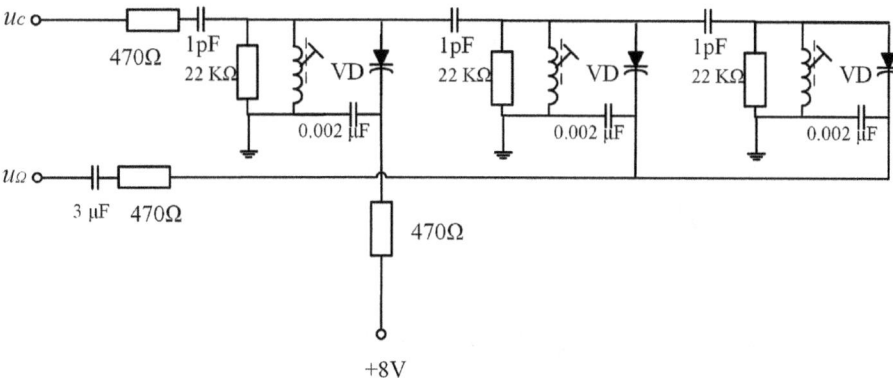

Fig. 5.15: Three-level PM circuit with varactors.

three 22 kΩ resistors, in order to obtain identical phase shift. Finally, a summed up phase shift within π/2 can be realized.

5.5 Demodulation of FM

As mentioned earlier, the information of the modulation signal has been loaded onto the instantaneous frequency or phase of the FM or PM wave. The inverse process of modulation in which the information of the modulation signal is decoded is called demodulation. In other words, this process is to complete frequency–voltage or phase–voltage conversion. The demodulation circuit is called frequency discriminator or phase detector.

5.5.1 Methods of frequency discrimination and the circuit model

5.5.1.1 Methods of frequency discrimination
There are essentially two kinds of thoughts for frequency discrimination, phase locked loop (PLL) or waveform conversion. PLL-based frequency discrimination will be discussed in Chapter 6 briefly. For the other waveform conversion, there are several detailed realization methods that we would like to introduce to the readers.

The first one is to convert FM waveform into FM–AM waveform by a frequency–amplitude network; then use envelope detection (amplitude demodulation) to extract the modulation signal out which is the same as what is loaded on the FM signal. This kind of frequency discrimination can be done by slope detectors with the principle shown in Fig. 5.16.

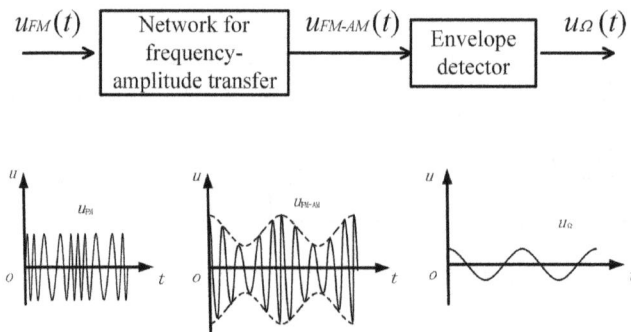

Fig. 5.16: Circuit principle for frequency–amplitude discriminator.

The second method is to convert FM waveform into FM–PM waveform by a frequency–phase network; then a phase detector is applied to get the modulation signal.

This method is based on phase discrimination with the circuit block diagram shown in Fig. 5.17.

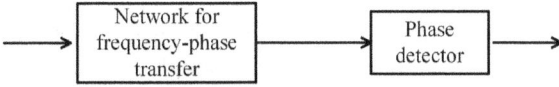

Fig. 5.17: Block diagram of phase discriminator.

The third method is to use frequency synthesis by analog multipliers. In this method, the FM signal is split into two paths. One path is added with a $\pi/2$ phase-shift, mixing with the other path which is the original FM signal by a multiplier. A low-pass filter afterwards collects the modulation signal out of several frequency components after multiplication operation. The circuit block diagram is shown in Fig. 5.18.

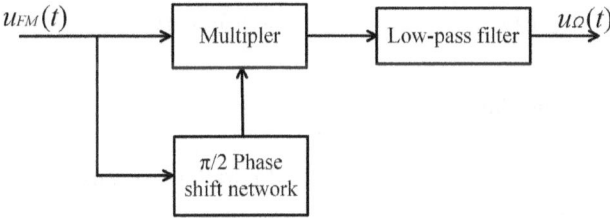

Fig. 5.18: Block diagram of FM discrimination with an analog multiplier.

The fourth method is called pulse counting discrimination (Fig. 5.19). The advantages of this method are good linearity and broadband frequency. However, the disadvantage of this method is that the carrier frequency is not high.

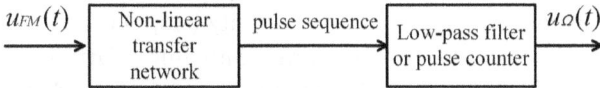

Fig. 5.19: Block diagram of pulse counting discrimination.

5.5.2 Performance indices of frequency discriminators

(1) Sensitivity g_d. Frequency discrimination sensitivity g_d is defined as the output voltage caused by unit frequency shift near the central frequency f_0.

$$g_d = \frac{du_\Omega}{df}\bigg|_{f=f_0} \approx \frac{\Delta u_\Omega}{\Delta f}\bigg|_{f=f_0} \tag{5.39}$$

(2) Bandwidth B_m. The frequency discrimination characteristics curve is close to linear within bandwidth B_m. The basic requirement on bandwidth is that B_m should be at least twice the FM signal's frequency shift.

(3) Nonlinear distortion. Within bandwidth B_m, output voltage u_Ω and frequency shift Δf is approximately linear, indicating the existence of nonlinear distortion. We hope that nonlinear distortion is as trivial as possible though. Fig. 5.20 is a frequency discrimination curve.

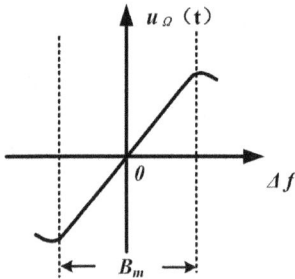

Fig. 5.20: Frequency discrimination characteristics.

5.5.3 Slope of frequency discrimination

Slope frequency discriminator is made up of the mistuned parallel resonance loop and envelope detection circuit (Fig. 5.21).

The resonance frequency of the circuit is set at the location which is slightly away from the center carrier frequency. Consequently, when FM signals enter the mistuned LC resonance circuit, the output voltage amplitude varies nearly proportionally with the actual frequency. The input FM's central frequency is kept at the center of the linear amplitude–frequency region to ensure the optimal performance. At the end of this process the FM signal with uniform amplitude is changed into a waveform whose amplitude varies coherently with the frequency.

In order to improve the linear region as well as sensitivity (slope) of the frequency discriminator, staggering mistuned frequency discriminator is further introduced. As shown in Fig. 5.22, two slope discriminators are connected side by side. The corresponding resonant frequencies are ω_{p1} and ω_{p2}, which are located at the two separate sides of the carrier frequency ω_c with equal deviations: $\Delta\omega_{c1} = \Delta\omega_{c2} = \omega_{p1} - \omega_c = \omega_c - \omega_{p2}$. We assume the amplitude-frequency characteristics for the two mistuned loops are $A_1(\omega)$ and $A_2(\omega)$, respectively. Then the total output of the circuit is

$$u_\Omega = \eta_d U_{FM}[A_1(\omega) - A_2(\omega)] \tag{5.40}$$

Where U_{FM} is the amplitude of the FM signal and η_d is the detection efficiency.

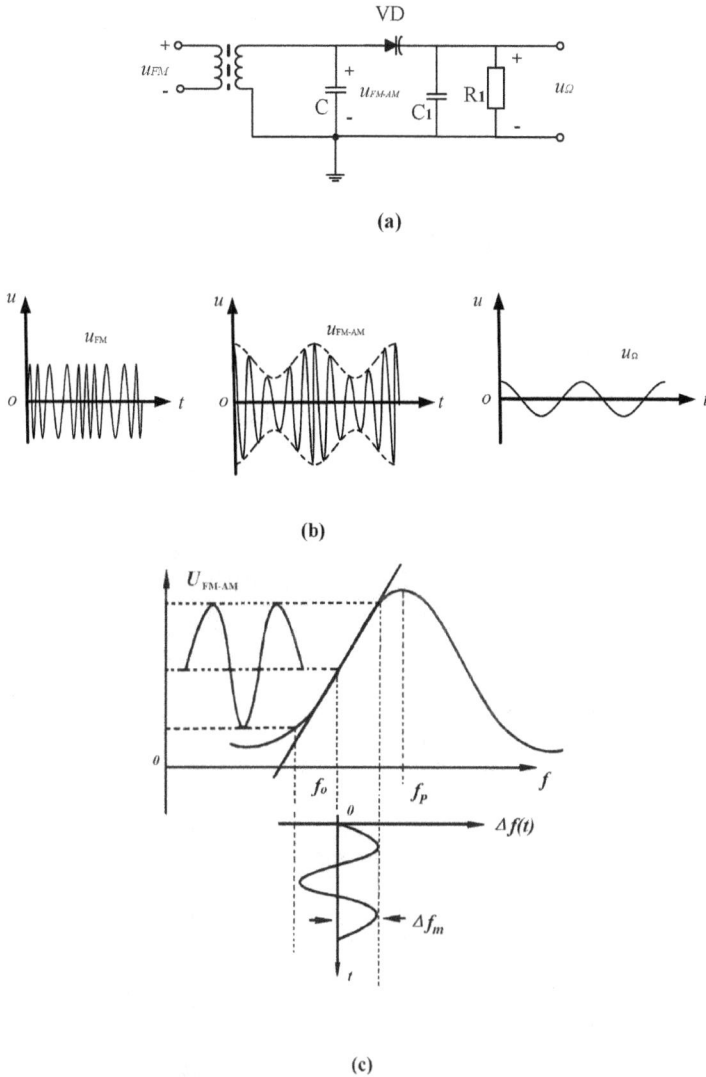

Fig. 5.21: Slope frequency discriminator, (a) circuit schematic, (b) waveforms, (c) function of amplitude versus frequency.

5.5.4 Phase discriminator

A phase discriminator contains a frequency–phase conversion network and a phase detector and phase detector is the key part.

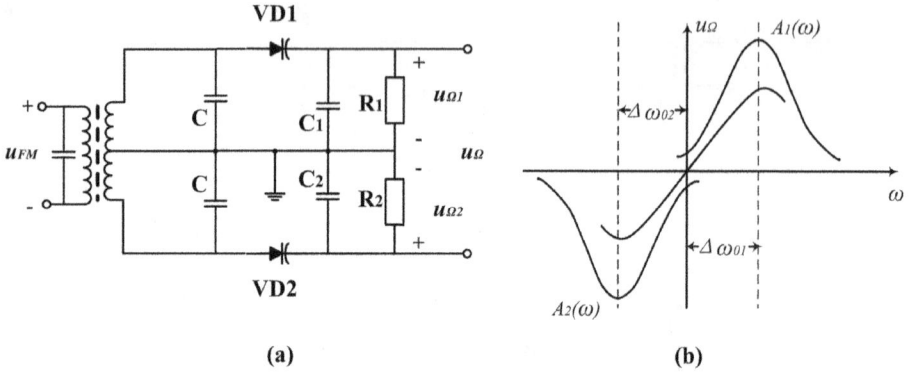

Fig. 5.22: Staggering mistuned frequency discriminator. (a) Circuit model, (b) amplitude-frequency characteristics.

5.5.4.1 Phase detector

The target of phase detector is to recognize the modulated wave's instantaneous phase and generate a proportional voltage signal according to the phase change. There are two kinds of phase detector.

5.5.4.1.1 Multiplier phase detector

The circuit block diagram is shown in Fig. 5.23.

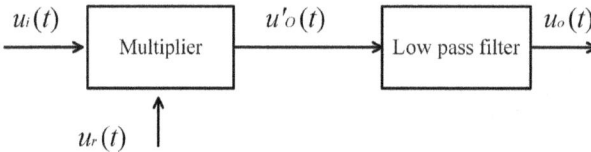

Fig. 5.23: Circuit block diagram of a multiplier phase detector.

Suppose the first input PM signal is $u_i = U_{im} \cos[\omega_c t + \varphi(t)]$, $\varphi(t) = k_p u_\Omega (t)$. And the second input signal is its orthogonal wave $u_r = U_{rm} \sin \omega_c t$. The output signal of the multiplier is

$$u'_o(t) = K u_i u_r = K U_{im} U_{rm} \cos[\omega_c t + \varphi(t)] \cos\left(\omega_c t + \frac{\pi}{2}\right)$$

$$= \frac{1}{2} K U_{im} U_{rm} \left\{ \cos\left[\varphi(t) - \frac{\pi}{2}\right] + \cos\left[2\omega_c t + \varphi(t) + \frac{\pi}{2}\right] \right\}$$

(5.41)

When the output signal is put into a low-pass filter, the low-frequency part is obtained. The final output is

$$u_o(t) = \frac{1}{2}KU_{im}U_{rm}\cos\left[\varphi(t) - \frac{\pi}{2}\right] = \frac{1}{2}KU_{im}U_{rm}\sin\varphi(t) \tag{5.42}$$

As $|\varphi(t)| \le \frac{\pi}{12}$ and $\sin\varphi(t) \approx \varphi(t)$, the output is approximately linear with the phase

$$u_o(t) = \frac{1}{2}KU_{im}U_{rm}\varphi(t) = \frac{1}{2}KU_{im}U_{rm}k_p u_\Omega(t) \propto \varphi(t) \tag{5.43}$$

From the above analysis we can see linear phase detection is realized.

5.5.4.1.2 Adder phase detector
The circuit block diagram is shown in Fig. 5.24.

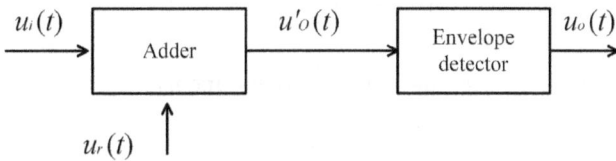

Fig. 5.24: Circuit block diagram of an adder phase detector.

A good example of the adder phase detector is shown in Fig. 5.25 which contains a balance structure.

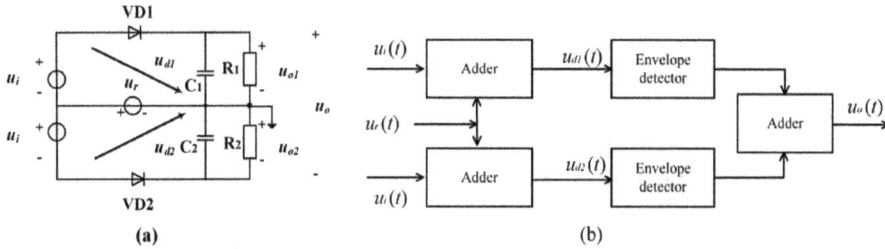

Fig. 5.25: Balance adder phase detector, (a) circuit model, (b) block diagram.

Suppose one of the input PM signals is

$$u_i(t) = U_{im}\cos[\omega_c t + \varphi(t)], \quad \varphi(t) = k_p u_\Omega(t)$$

Another signal is its orthogonal form $u_r(t) = U_{rm} \cos\left(\omega_c t + \dfrac{\pi}{2}\right)$. From the circuit we can get

$$\begin{cases} u_{d1} = u_r(t) + u_i(t) \\ u_{d2} = u_r(t) - u_i(t) \end{cases} \tag{5.44}$$

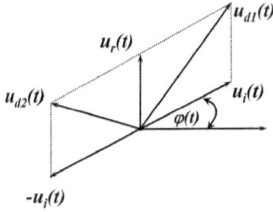

Fig. 5.26: Vector graph of voltage.

According to the vector graph in Fig. 5.26, the amplitude of u_{d1} and u_{d2}, is

$$\begin{cases} U_{d1} = \sqrt{U_i^2 + U_r^2 + 2U_iU_r \sin \varphi(t)} \\ U_{d2} = \sqrt{U_i^2 + U_r^2 - 2U_iU_r \sin \varphi(t)} \end{cases} \tag{5.45}$$

Apparently the amplitude of u_{d1} and u_{d2} is changing with the instantaneous phase $\varphi(t) = k_p u_\Omega(t)$. Both signals are PM–AM signal.

Suppose the envelope detector efficiencies are $\eta_{d1} = \eta_{d1} = \eta_d$, hence the output voltage of the up and down envelope detector can be expressed as

$$\begin{cases} u_{o1} = \eta_d U_{d1} \\ u_{o2} = \eta_d U_{d2} \end{cases} \tag{5.46}$$

The total output voltage is $u_o(t) = u_{o1} - u_{o2} = \eta_d(U_{d1} - U_{d2})$.

(1) When $U_i \ll U_r$, from eq. (5.45),

$$U_{d1} = U_r \sqrt{\left(\frac{U_i}{U_r}\right)^2 + 1 + 2\frac{U_i}{U_r}\sin \varphi(t)} \approx U_r \sqrt{1 + 2\frac{U_i}{U_r}\sin \varphi(t)}$$

$$\approx U_r \left[1 + \frac{U_i}{U_r}\sin \varphi(t)\right] \tag{5.47}$$

Similarly,

$$U_{d2} \approx U_r \left[1 - \frac{U_i}{U_r}\sin \varphi(t)\right] \tag{5.48}$$

Then $u_o(t) = u_{o1} - u_{o2} = 2\eta_d(U_{d1} - U_{d2}) = 2\eta_d U_i \sin \varphi(t)$. When $|\varphi(t)| \leq \dfrac{\pi}{12}$,

$$u_o(t) \approx 2\eta_d U_i \varphi(t) \tag{5.49}$$

It is a linear phase discriminator.

(2) When $U_i \gg U_r$,

$$u_o(t) = 2\eta_d U_r \sin \varphi(t) \approx 2\eta_d U_r \varphi(t) \tag{5.50}$$

(3) When $U_i = U_r = U$, according to eq. (5.45),

$$u_o(t) = u_{o1} - u_{o2} = 2\eta_d[U_{d1} - U_{d2}] = 2K_d U \sin \varphi(t) \approx 2\eta_d U \varphi(t) \tag{5.51}$$

The third case is usually used in an inductive coupling phase discriminator. Its circuit is shown in Fig. 5.27.

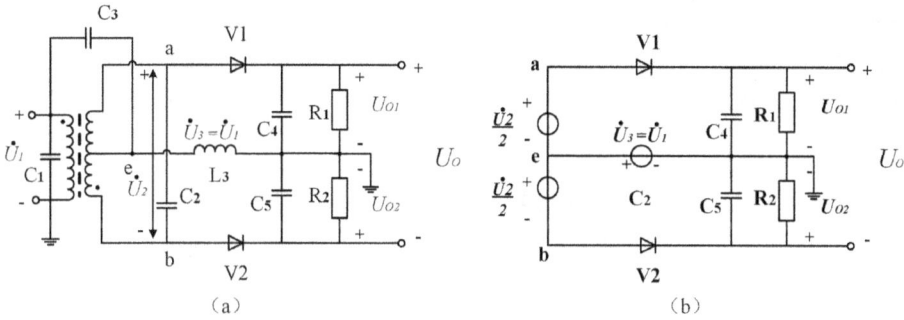

Fig. 5.27: Inductive coupling phase discriminator, (a) schematic circuit, (b) AC equivalent circuit.

In the above circuit, L_1C_1 and L_2C_2 are loosely-coupled double-resonance loops. Both loops resonate at ω_c. When the input frequency changes the output signal of the secondary resonance loops cannot keep pace with the input signal. Hence the output phase differs from the input. As analyzed earlier, the output voltage would change accordingly too. Finally, this circuit is capable of converting frequency change to an amplitude change.

There are two pathways of signal coupling. The first one is through the inductive coil. It is obtained at the two ends of L_2C_2. Point e is the center of coil L_2, and a-e and b-e have the same voltage, which is equal to $\frac{\dot{U}_2}{2}$. The second path is through coupling capacitor C_3. L_3 being a high-frequency chock, its high-frequency resistance is much greater than C_3. Voltage $\dot{U}_3$ obtained from the two ends of L_3 is nearly equal to voltage $\dot{U}_1$ on the primary loop. $\dot{U}_3$ provides DC bias for the main circuit. The entire circuit could be simplified as shown in Fig. 5.27(b), which is similar with the circuit of the adder phase detector. Compared with previous analysis, we can obtain the voltage on the two diodes

$$\begin{cases} \dot{U}_{d1} = \dot{U}_3 + \dfrac{\dot{U}_2}{2} \approx \dot{U}_1 + \dfrac{\dot{U}_2}{2} \\[2mm] \dot{U}_{d2} = \dot{U}_3 - \dfrac{\dot{U}_2}{2} \approx \dot{U}_1 - \dfrac{\dot{U}_2}{2} \end{cases} \tag{5.52}$$

The output voltage of the top and bottom loop is as follows:

$$\begin{cases} u_{o1} = \eta_d U_{d1} \\ u_{o2} = \eta_d U_{d2} \end{cases} \tag{5.53}$$

The total output voltage of the circuit is

$$u_o = u_{o1} - u_{o2} = \eta_d (U_{d1} - U_{d2}) \tag{5.54}$$

The output voltage changes according to the input frequency. We can comb our thought by clearing up the relationships between different variables step by step: (1) the phase difference that $\dot{U}_2$ falls behind $\dot{U}_1$ is a variable, which changes with the primary frequency; (2) the voltage on the two diodes, that is, the envelope detector's input voltage, U_{d1} and U_{d2} change with the input frequency; (3) output voltage u_{o1}, u_{o2} of the envelope detector change with the input frequency; and (4) the output voltage of the entire circuit change with the input frequency.

Let us analyze in detail the above relationships.

(1) Phase difference between the primary and the secondary.

In Fig. 5.27, the current going through L_1 is

$$\dot{I}_1 = \frac{\dot{U}_1}{R_1 + j\omega L_1 + \frac{(\omega M)^2}{Z_2}} \tag{5.55}$$

where R_1 and L_1 are the resistance and inductance at the primary circuit, M is the mutual inductance and Z_2 is the impedance at the secondary circuit. Suppose Q factor of the resonance loop is high enough, then the loss of the primary inductance and the secondary back onto the primary side can be ignored. And we have the current going through L_1 to be

$$\dot{I}_1 \approx \frac{\dot{U}_1}{j\omega L_1} \tag{5.56}$$

The instantaneous phase of $\dot{I}_1$ lags behind $\dot{U}_1$ for $\frac{\pi}{2}$, which is shown in Fig. 5.28.

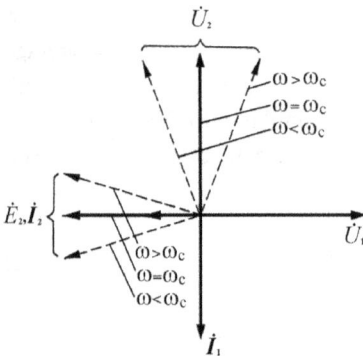

Fig. 5.28: Vector diagram of $\dot{U}_2$, $\dot{I}_2$ and $\dot{E}_2$.

$\dot{I}_1$ induces the electromotive voltage in the secondary circuit due to the mutual inductance M:

$$\dot{E}_2 = -j\omega M \dot{I}_1 \tag{5.57}$$

The instantaneous phase of $\dot{E}_2$ lags behind $\dot{I}_1$ for $\dfrac{\pi}{2}$.
$\dot{E}_2$ causes current $\dot{I}_2$

$$\dot{I}_2 = \frac{\dot{E}_2}{Z_2} = \frac{\dot{E}_2}{R_2 + j\left(\omega L_2 - \frac{1}{\omega C_2}\right)} \tag{5.58}$$

where R_2 is the resistance of the secondary side. According to eq. (5.58), the instantaneous phase of $\dot{I}_2$ changes with the primary frequency.

① When $\omega = \omega_c$, the circuit is in resonance and $\dot{I}_2$ is in phase with $\dot{E}_2$.
② When $\omega > \omega_c$, the circuits is inductive and the instantaneous phase of $\dot{I}_2$ lags behind $\dot{E}_2$.
③ When $\omega < \omega_c$, the circuits is capacitive, the instantaneous phase of $\dot{I}_2$ is ahead of $\dot{E}_2$.

$\dot{U}_2$, defined as the voltage across C_2, lags behind $\dot{I}_2$ for $\frac{\pi}{2}$.

$$\dot{U}_2 = \dot{I}_2 \cdot \frac{1}{j\omega C_2} = -\frac{j\omega M \dot{I}_1}{R_2 + j\left(\omega L_2 - \frac{1}{\omega C_2}\right)} \frac{1}{j\omega C_2}$$

$$= j\frac{1}{\omega C_2}\frac{M}{L_1}\frac{\dot{U}_1}{R_2 + j\left(\omega L_2 - \frac{1}{\omega C_2}\right)} \tag{5.59}$$

According to eq. (5.59), the instantaneous phase of the secondary voltage $\dot{U}_2$ changes with the primary frequency.

① When $\omega = \omega_c$, the instantaneous phase of $\dot{U}_2$ leads $\dot{U}_1$ for $\frac{\pi}{2}$.
② When $\omega > \omega_c$, the instantaneous phase of $\dot{U}_2$ leads U_1 for less than $\frac{\pi}{2}$.
③ When $\omega < \omega_c$, the instantaneous phase of $\dot{U}_2$ leads $\dot{U}_1$ for more than $\frac{\pi}{2}$.

(2) U_{d1} and U_{d2}, the detector's amplitude, changes with the primary frequency.

According to $\begin{cases} \dot{U}_{d1} \approx \dot{U}_1 + \frac{\dot{U}_2}{2} \\ \dot{U}_{d2} \approx \dot{U}_1 - \frac{\dot{U}_2}{2} \end{cases}$, the vector diagram is shown in Fig. 5.29.

① When $\omega = \omega_c$, $U_{d1} = U_{d2}$.
② When $\omega > \omega_c$, $U_{d1} > U_{d2}$.
③ When $\omega < \omega_c$, $U_{d1} < U_{d2}$.

(3) u_{o1} and u_{o2}, the output voltage of the detector, changes with the input frequency. Because $u_{o1} = \eta_d U_{d1}$, $u_{o2} = \eta_d U_{d2}$. They are all proportional to U_d.
(4) u_o, the final output voltage of the circuit, changes with the input frequency.

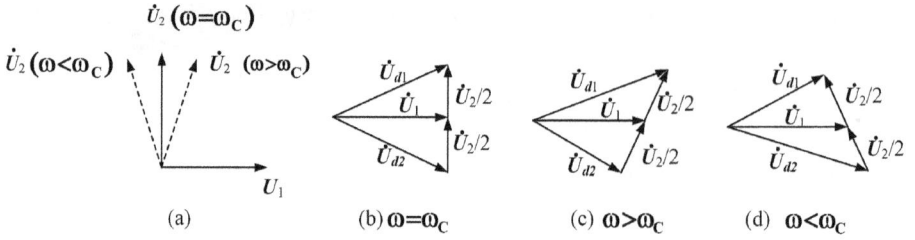

Fig. 5.29: Vector diagram of $\dot{U}_1$ and $\dot{U}_2$, (a) U2 with ω, (b) $\omega = \omega_c$, (c) $\omega > \omega_c$, (d) $\omega < \omega_c$.

The final output can be written as $u_o = u_{o1} - u_{o2}$.

① When $\omega = \omega_c$, $u_{o1} = u_{o2}$ and $u_o = 0$.

② When $\omega > \omega_c$, $u_{o1} > u_{o2}$ and $u_o r > 0$.

③ When $\omega < \omega_c$, $u_{o1} < u_{o2}$ and $u_o < 0$.

The $u_o - \omega(\Delta f)$ curve is shown in Fig. 5.30.

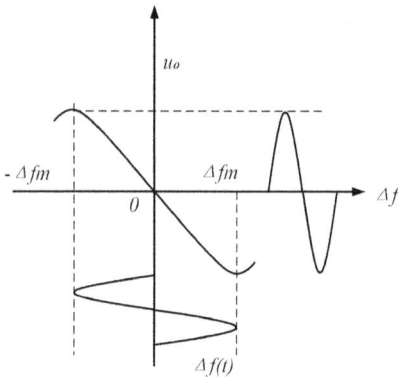

Fig. 5.30: Characteristic curve of the discriminator.

In conclusion, the double-loop inductive coupling phase discriminator is a frequency–phase conversion network. It converts FM wave $u_1(t)$ into an FM–PM signal $u_2(t)$. After super position of two signals, two FM–AM signals, $u_{d1}(t)$ and $u_{d2}(t)$ are obtained. Finally, the FM–AM signal is detected by envelope detector to get the modulation signal extracted.

By adjusting the coupling coefficient $k = \dfrac{M}{\sqrt{L_1 L_2}}$ and the quality factor Q, the $u_o - \Delta f$ curve can be carefully tuned to obtain optimal linearity of the demodulation and sufficient bandwidth B.

.5.5 Ratio discriminator

Ratio discriminator looks very similar with the discussed inductive coupling phase discriminator and it does not need a limiter. There are several differences in a ratio discriminator circuit worth noting.
(1) Diode V_1 is connected reversely.
(2) A large capacitor C_5 (10 µF) is added across R_3 and R_4.
(3) Points M and E are not electrically connected any more. The output voltage is obtained between M and E (instead of between F and G in the previous case). Current through C_3 and C_4 is with opposite direction, which results in the output to be a differential voltage.

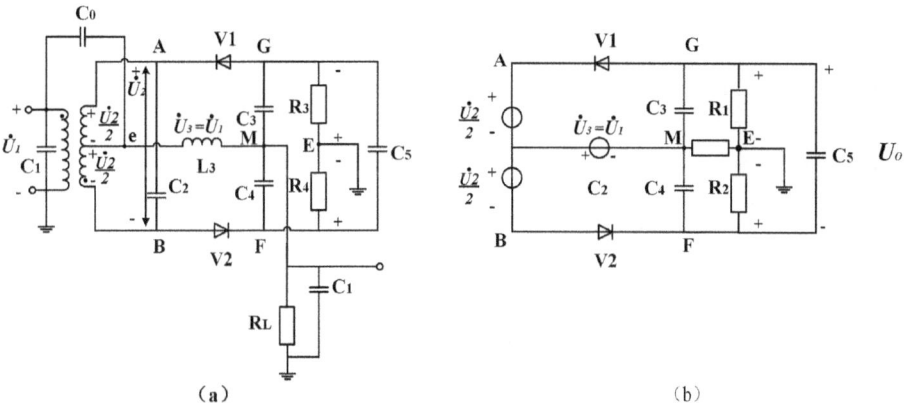

Fig. 5.31: Ratio discriminator, (a) schematic circuit, (b) AC equivalent circuit.

Let us analyze how this circuit functions.

According to the circuit diagram, voltage drop on C_3 and C_4 will be U_{o1} and U_{o2} after the envelope detector. The voltage drop on C_5 then should be $U_c = U_{o1} + U_{o2}$.

C_5 is a very large capacitor, meaning its discharging time constant $C_5(R_3 + R_4)$ is long (roughly 0.1–0.2 s). In an audio cycle, the voltage of C_5 can be nearly treated as a constant which does not change.

Because $R_3 = R_4$, the voltage on each of them is half of U_C. We can get the voltage of points F and G as

$$\begin{cases} U_F = \dfrac{U_C}{2} \\ U_G = -\dfrac{U_C}{2} \end{cases} \tag{5.60}$$

Similar with earlier circuits, when the input frequency ω changes the voltage on C_3 and C_4 changes too. Since the voltage of points F and G is nearly fixed, U_M has to change.

Therefore, we can get the following analysis.
(1) When $\omega = \omega_c$, $U_{o1} = U_{o2}$ and $U_M = 0$.
(2) When $\omega > \omega_c$, $U_{o1} > U_{o2}$ and $U_M \uparrow$.
(3) When $\omega < \omega_c$, $U_{o1} < U_{o2}$ and $U_M \downarrow$.

U_M reflects the change of ω. The voltage at points F, G and M is shown in Fig. 5.32.

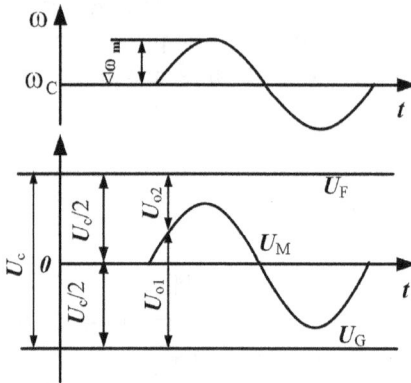

Fig. 5.32: Potential of points F, G and M.

One of ratio discriminator's biggest advantages is the ability to limit amplitude variation automatically, which mainly attributes to the large capacitor C_5. Because of C_5, the voltage on C_3 and C_4 together is almost a constant of U_C. Suppose the amplitude of the high-frequency signal increases unexpectedly, the existence of C_5 will absorb most of the charges keeping the voltage of $C_3 + C_4$ stable. That suggests within one cycle, the charging time and current will increase and the detector will consume more power. The power being supplied by the resonance circuit, the Q of the resonance circuit decreases. Then the voltage of the resonance circuit decreases. That compensates the increasing of the signal. Similarly the signal's sudden decrease of amplitude will also be compensated.

The adjustment of Q of the resonance circuit makes ratio discriminator has the function of a limiter, which makes the entire circuit more concise.

5.5.6 Pulse-counting discriminator

The working principle of this FM discriminator is different from those introduced before. It is called pulse-counting discriminator because it counts the number of the zero-crossing pulses to read frequency information directly. For its advantage of good linearity and wide bandwidth, pulse-counting discriminators are used broadly and frequently integrated into IC.

The basic working principle of pulse-counting discriminator is to transfer the FM into a pulse sequence with constant amplitude and width which has the same repetitive frequency with instantaneous frequency of the FM signal. Consequently, the average DC component is extracted which is the modulation signal embedded. The block diagram and waveform are shown in Figs. 5.33 and 5.34.

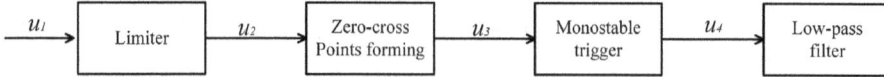

Fig. 5.33: Circuit block diagram of the pulse-counting discriminators.

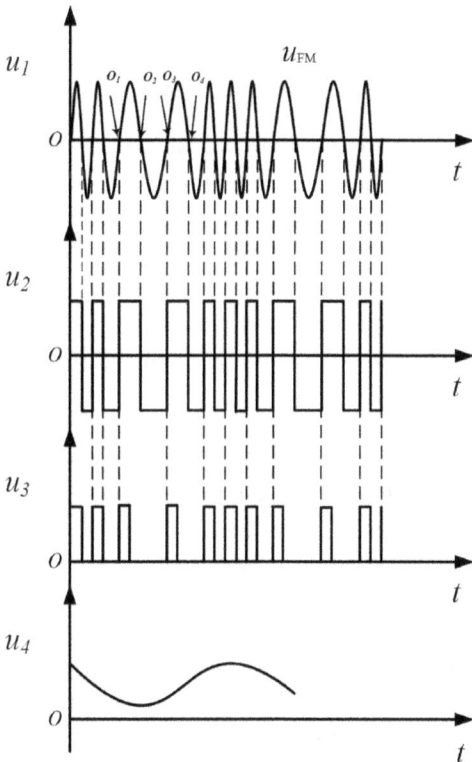

Fig. 5.34: Waveforms of a pulse-counting discriminator.

An FM signal u_1 is sent through a limiter and put into the zero-forming class to obtain cross-zero points or pulse sequence u_2. This process can be realized with a Schmidt circuit. Pulse sequence u_2 with constant amplitude but different width is then applied to trigger a monostable multivibrator. The trigger uses the rising edge. Finally a pulse sequence u_2 with constant amplitude and width is realized.

The number of pulses in sequence u_2 counts the frequency of u_1. To understand this correlation, first of all we need to know that the density of cross-zero points are determined by the vibrations per unit time which is frequency. In Fig. 5.34, O_1, O_2 and O_3 . . . are all zero-cross points. O_1, O_3, . . ., from positive to negative, are called positive zero-cross points. And O_2, O_4, . . . are called negative zero-cross points. Therefore, frequency information of u_1 can be counted by the positive zero-cross points. The repetitive frequency of the rectangular pulses u_3 reflects the instantaneous frequency of u_1. The DC component (or average amplitude) of u_3 waveform stands for the low FM signal which was embedded in the FM signal u_1. After u_3 passes a low-pass filter, the modulation signal is successfully extracted.

5.6 Limiter

5.6.1 Introduction

The modulation signal can be loaded with parasitic amplitude in the process of signal transmission by various reasons. Many frequency discriminators (except for ratio frequency discriminator) take the parasitic amplitude information all the way to the output voltage causing unwanted distortions and variations, which lead to poor communication quality. In order to eliminate the effect of parasitic amplitude, the first-class limiter can be added before a frequency discriminator. The expectation for a limiter is to remove the parasitic amplitude without changing the frequency information of the FM signal.

Amplitude limiting is a nonlinear process and new frequencies may be generated during the process. So, the limiter usually consists of nonlinear device and resonant tank. When an FM signal with parasitic amplitude goes through the nonlinear device, the wave form is cut flat at its maximums and minimums and therefore distorted somehow. Behind the limiter is a resonant tank or band-pass filter, which blocks the unnecessary frequencies and obtains the normal FM signal.

The characteristic of a limiter is showed in Fig.5.35. The curve explains the relationship between output voltage u_o and the input voltage u_i. At curve OA, the output increases with the increase of the input. After point A, the output almost stays constant despite the increase of input. Point A is called the clipping threshold, and the corresponding input voltage U_i is called the voltage threshold. Obviously only when the input voltage exceeds the voltage threshold U_i, amplitude can be limited effectively.

There are different kinds of limiter circuits. In this chapter we will introduce two examples: diode limiter and transistor limiter.

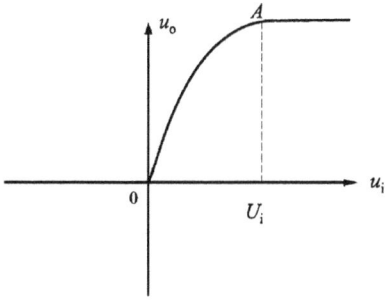

Fig. 5.35: Characteristic of a limiter.

5.6.2 Diode limiter

A kind of diode limiter is shown in Fig. 5.36. Comparing with a common FM circuit, a pair of forward and reverse diodes (D_1 and D_2) is placed in parallel in the resonance circuit. When the input signal is small, the resonance loop's voltage is low. If the loop voltage is lower that the diode's threshold voltage, the diodes are off and do not affect the circuit. When the input signal is large enough, two diodes will be switched on alternately. The forward resistance of diodes changes with the forward voltage. Higher input voltage brings lower resistance, lower quality factor and hence smaller output voltage. The output voltage is limited at the level of the threshold voltage of the diode [20].

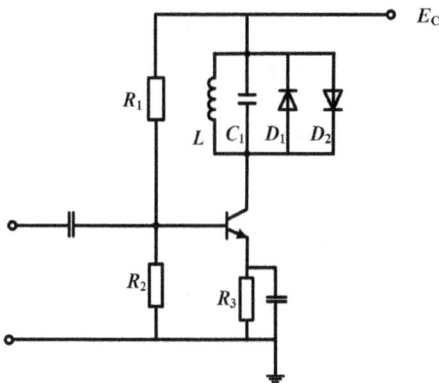

Fig. 5.36: Diode limiter.

5.6.3 Transistor limiter

Transistor limiters use transistor as the clipping device (Fig. 5.37). Different and special operation condition is employed here compared with other transistor applications. If the input signal is small, the limiter circuit works as a normal intermediate frequency amplifier; if the input increases to a certain extent, the limiter circuit works in the

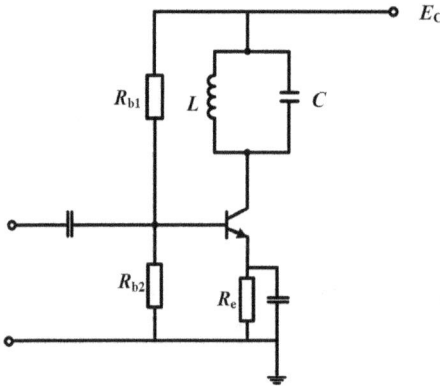

Fig. 5.37: Transistor limiter circuit diagram.

saturation or cutoff region. When saturation and cutoff durations are equal, the circuit function of amplitude limiting can be realized.

A conventional limiting circuit cannot keep the output voltage amplitude strictly constant but only limit it to a certain degree. The limiting function is not valid at all especially when the signal is weak enough. Therefore, a good limiter is required to be high signal gain and not only limit at one class.

5.7 Comparison of different modulation methods

AM, FM and PM have different properties. We would like to compare these three modulation methods from multiple aspects in the following content.

5.7.1 Antiinterference ability

The distance and reliability of communication highly relies on the signal's antiinterference ability. If there is no interference or the interference has little effect on the signal, the communication distance can be very long even if the transmission power is weak. In fact the interference exists everywhere, so the antiinterference ability is a very important performance index. Generally, the antiinterference ability of FM systems is better than AM systems. But when the ratio between the receiving signal and the interference intensity (r_1) is lower than a certain critical value, the FM systems may perform worse than AM systems. Figure 5.38 shows the change of ratio between the output signal and the interference (r_2) under different r_1. We can compare the solid curves (FM) with the dashed line (AM) for the antiinterference ability. Under the same r_1 value, a higher r_2 is considered superior in antiinterference. For the case of $\frac{\Delta\omega_f}{\Omega_{max}} = 1$, the critical value of r_1 is about 4 dB. Beyond $r_1 = 4$ dB, FM is better than AM

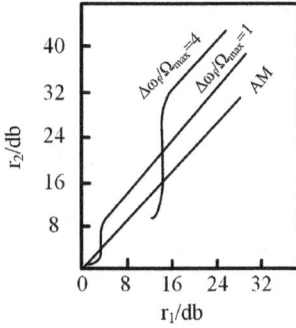

Fig. 5.38: The influence of the receiving signal/interference ratio (r_1) on the output/interference ratio (r_2) for FM signals.

and under this value, AM is better than FM. For the case of $\frac{\Delta\omega_f}{\Omega_{max}} = 4$, the critical value of r_1 is roughly 16 dB. Through this comparison we observe that antiinterference ability of FM is significantly better than AM only when the received signal is much stronger than the interference. Moreover, larger $\Delta\omega_f$ requires larger r_1 to be good at antiinterference. Therefore, FM with large frequency shift (wide-band FM) only beats weak interference. And the FM with small frequency shift (narrow-band FM) works with interference with intermediate intensity.

5.7.2 Bandwidth occupation

FM needs to occupy wider bandwidth than AM, but with better communication quality as trade off.

5.7.3 Transmitter power requirement

Because the constant amplitude of FM signals, the power consumption does not change with different modulation conditions. However, the power of AM changes along with the modulation depth. Typically, FM needs lower power from transmitters than AM.

5.7.4 Strong signal blockage

In communication systems, signal intensity differs greatly for different transmission distance. When the signal is very strong, receivers usually work at the amplitude limiting state. That brings serious distortion or even the failure to AM signal receiving. FM could avoid this blockage caused by strong signal to some extent.

5.8 Low-power FM receiver system based on digital IC

With the advance of IC, it becomes the trend that the whole communication system be realized on one chip. IC chips, such as MC2833, MC 3362 and so on, are convenient to accomplish low power FM transmitting and receiving directly. As an example, receiving system based on MC3362 will be introduced.

MC3362 (Fig. 5.39) is a product of Motorola featuring low noise, low power consumption (low voltage: V_{CC} = 2.0–6.0 V_{dc}, low drain current = 3.6 mA typical at V_{CC} = 3.0 V_{dc}) and excellent sensitivity (input voltage 0.6 mV rms for 12 dB SINAD). The powerful internal circuits allow its good performance with a few external devices. Its pin connections and representative block diagram are drawn in Fig. 5.40 [21].

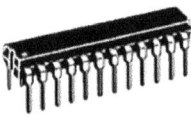

P SUFFIX
PLASITIC PACKAGE
CASE 724

DW SUFFIX
PLASITIC PACKAGE
CASE 751E

Fig. 5.39: MC3362 chips (with two types of packages).

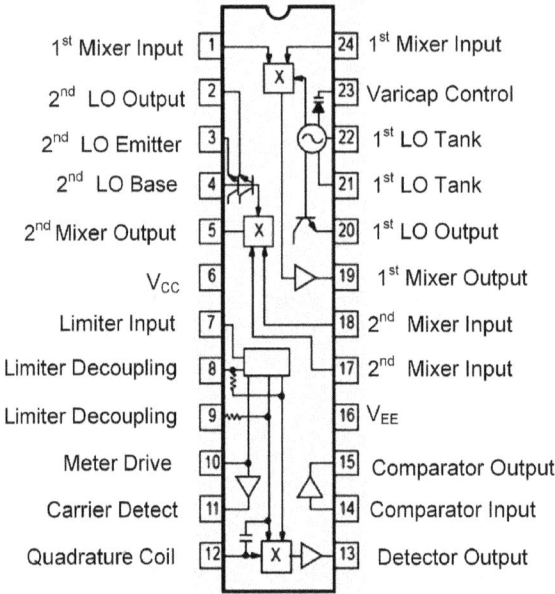

Pin	Left		Right	Pin
1	1st Mixer Input		1st Mixer Input	24
2	2nd LO Output		Varicap Control	23
3	2nd LO Emitter		1st LO Tank	22
4	2nd LO Base		1st LO Tank	21
5	2nd Mixer Output		1st LO Output	20
6	V_{CC}		1st Mixer Output	19
7	Limiter Input		2nd Mixer Input	18
8	Limiter Decoupling		2nd Mixer Input	17
9	Limiter Decoupling		V_{EE}	16
10	Meter Drive		Comparator Output	15
11	Carrier Detect		Comparator Input	14
12	Quadrature Coil		Detector Output	13

Fig. 5.40: Pin connections and representative block diagram.

The receiver based on MC3362 is shown in Fig. 5.41 [21]. It includes dual FM conversion with oscillators, mixers, quadrature discriminator and meter drive/carrier detect circuitry. The MC3362 also has buffered first and second local oscillator outputs and a comparator circuit for FSK detection.

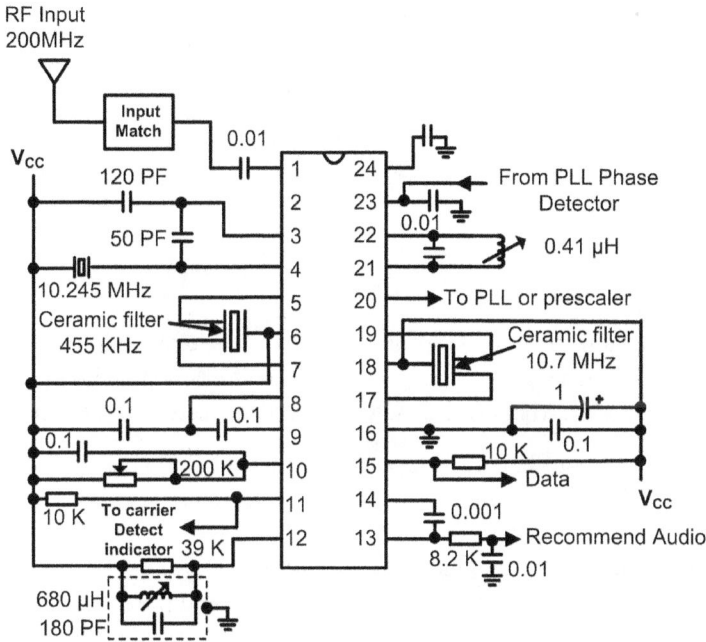

Fig. 5.41: An integrated FM receiver based on MC3362.

Problems

5.1 The expression of an angular modulation signal is $u(t) = 10\cos(2\pi \times 10^6 t + 10\cos2\pi \times 10^3 t)$, judge if it is PM or FM, and then calculate its
(1) maximum frequency shift;
(2) maximum phase shift;
(3) bandwidth;
(4) power on a unit resistance.

5.2 A carrier wave is given as $f_c = 100$ MHz and $U_{cm} = 5$ V. The modulation signal has the expression of $u_\Omega(t) = \cos2\pi \times 10^3 t + 2\cos2\pi \times 500 t$. Write the expression of the FM wave, assuming that the frequency shift for both parts of the modulation signal is $\Delta f_{max} = 20$ kHz.

5.3 An AM and an FM wave are generated by the same carrier wave and modulation signal with $f_c = 1$ MHz and $u_\Omega(t) = 0.1\sin2\pi \times 10^3\, t$. It is also known that for FM signal, the frequency shift caused by unit modulation voltage is 1 kHz/V.
   (1) Calculate B_{AM}, the bandwidth of the AM wave and B_{FM}, the effective bandwidth of the FM wave.
   (2) When the modulation signal is changed to $u_\Omega(t) = 20\sin2\pi \times 10^3\, t$, recalculate B_{AM} and B_{FM}.

5.4 Carrier wave frequency $f_c = 100$ MHz and maximum frequency shift $\Delta f_{max} = 75$ kHz are known. Calculate the modulation factor and the effective B_{FM} given the different values of the sinusoidal modulation frequency F:
   (1) $F = 300$ Hz;
   (2) $F = 3$ kHz;
   (3) $F = 15$ kHz.

5.5 For a modulation signal with $F = 400$ Hz and $U_{\Omega m} = 2.4$ V, the modulation factor is 60. When the modulation signal's frequency is reduced to 250 Hz and the amplitude is increased to 3.2 V, what is the updated value of the modulation factor?

5.6 The frequency discrimination characteristics curve is shown in Ex.Fig. 5.1. The output voltage of the frequency discriminator is known to be $u_o(t) = \cos4\pi \times 10^3\, t$ (V). Calculate
   (1) the trans-conductance g_d of frequency demodulation.
   (2) the mathematical expression of the input signal $u_{FM}(t)$ and the modulation signal $u_\Omega(t)$.

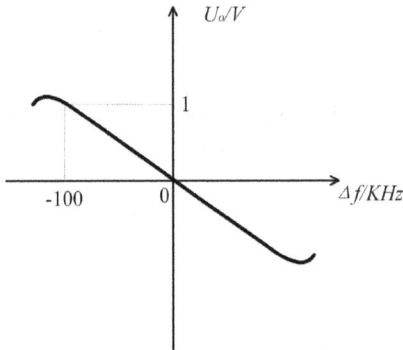

Ex.Fig. 5.1

5.7 Known that the carrier signal is $u_c(t) = U_{cm} \cos \omega_c t$ and the modulation signal is
$u_\Omega(t) = U_{\Omega m} \cos \Omega t$
  (1) Judge what type of signal $u_o(t) = U_{cm} \cos(\omega_c t + k_f \int_0^t u_\Omega(t)dt)$ is. (Tip: FM or PM.)
  (2) Draw the waveform of $u_o(t)$.
  (3) Draw the frequency spectrum of $u_o(t)$.
  (4) Write the bandwidth expression of $u_o(t)$.

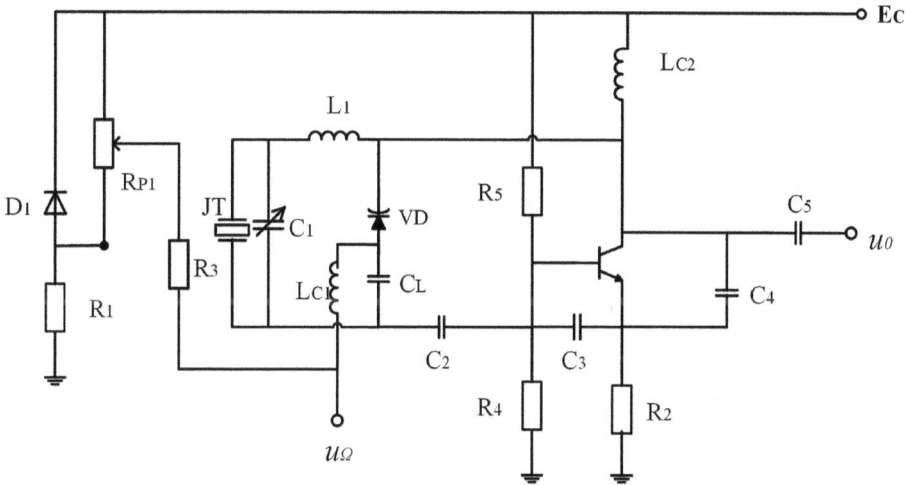

Ex.Fig. 5.2

5.8 A direct FM circuit with crystal oscillator is shown in Ex.Fig. 5.2. Explain the circuit principle and the function of the main devices.

5.9 For a frequency discriminator, the output wave is sinusoidal within its working range of $B_m = 2$ MHz. The input signal is $u_\Omega(t) = U_i \sin(\omega_c t + m_f \cos Ft)$. Calculate the output voltage under the following conditions: (1) $F = 1$ MHz, $m_f = 8$; (2) $F = 1$ MHz, $m_f = 10$.

5.10 Why the ratio frequency discriminator has the function of amplitude limiting?

5.11 Compare the three major signal modulation techniques of AM, FM and PM.

Chapter 6
Phase lock loop

6.1 PLL introduction

A phase lock loop (PLL) is a closed-loop feedback circuit that forces the phase of a voltage-controlled oscillator (VCO) to follow the phase of a stable reference signal. Once the lock is achieved, the frequency of the VCO will be exactly equal to the reference with a constant phase difference. The VCO is able to track the frequency of the reference signal over a certain operating region. Therefore an important class of applications of a PLL is a narrow band pass filter that passes a carrier while rejecting its noise sidebands. Early examples include the telemetry receivers which can track weak signals from spacecrafts despite the noise in space or the frequency shifts due to motion. Nowadays, PLL circuits are widely used such as PLL receiver, PLL oscillator, PLL-based frequency modulator and frequency modulation (FM) detector, PLL frequency multiplier, divider, mixer and so on. A PLL is capable of frequency synthesis as well as narrow band pass filtering with self-adjustable central frequency, which made PLL circuits irreplaceable in communication systems.

6.2 Basic components and principle

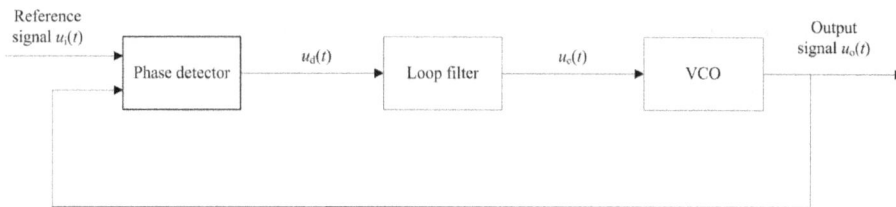

Fig. 6.1: PLL operational principle.

In the basic form, there are three fundamental components in a PLL circuit – a phase detector (PD), a loop filter (LF) and a VCO as shown in Fig. 6.1. These three parts are arranged into a feedback loop and operates as follows. A PD is used to compare the phase difference between the reference input signal u_i and the output signal of VCO u_o. The phase difference will be reflected in the output voltage u_d of the PD. A LF removes high-frequency components and noises of u_d and generates a control voltage u_c for VCO. The frequency of the output of VCO u_o will be adjusted

https://doi.org/10.1515/9783110593822-006

by u_c until u_o and u_i have exactly the same frequency, a constant phase difference and an invariable u_c. Then the lock status is achieved.

The whole system is considered linear in its operating region, which means all outputs and inputs hold a linear relation. In the subsequent contents we are going to analyze the linear model of each component.

6.3 Linear model

6.3.1 Phase detector model

A PD is a circuit capable of delivering a DC output voltage u_d that is proportional to the phase difference between the input phase θ_{in} and the VCO output phase θ_{out} (i.e., the PLL output phase).

$$u_d = K_{PD} \cdot (\theta_{in} - \theta_{out}) \qquad (6.1)$$

$$\theta_{err} \equiv \theta_{in} - \theta_{out} \qquad (6.2)$$

where K_{PD} is the proportionality factor and θ_{err} is defined as the phase difference. The output of PD u_d is later sent into VCO to affect its oscillation frequency [20].

For VCOs, the biasing voltage of the nominal frequency (free-running frequency) is usually set at a nonzero DC value, shown in Fig. 6.2, as V_0. So the effective voltage fed into VCO is

$$u_{err} = K_{PD} \cdot (\theta_{in} - \theta_{out}) + V_0 \qquad (6.3)$$

which is synthesized by the signal flow shown in Fig. 6.2. Figure 6.3 illustrates the ideal characteristic of PD.

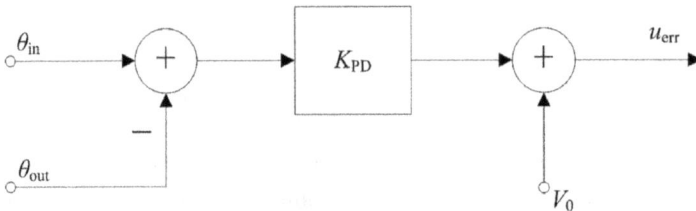

Fig. 6.2: Phase detector block diagram.

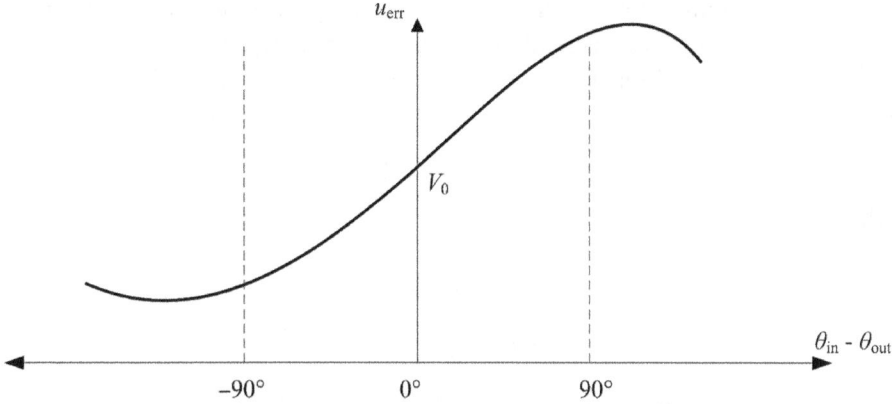

Fig. 6.3: Phase detector characteristic diagram.

There are multiple types of PDs, of which a multiplier (mixer) is widely used especially at high frequencies. Any ideal multiplier can be considered as a sine wave PD. To understand the principle and linearity of a PD we will conduct a complete deduction step by step. Assume that the form of the input signal u_i is

$$u_i\,(t) = U_{1m}\sin[\omega_i t + \theta_i(t)] \tag{6.4}$$

The output signal of VCO is

$$u_o(t) = U_{2m}\sin[\omega_o t + \theta_o(t)] \tag{6.5}$$

where U_{1m} and ω_i are the amplitude and angular frequency of the input signal; and U_{2m} and ω_o are amplitude and angular frequency of the VCO output signal, respectively. Unless the PLL stays locked, ω_i and ω_o are not necessarily identical. To see from the aspect of phase, we will rewrite the above expression of u_i and u_o in the form of their instantaneous phase.

$$u_i(t) = U_{1m}\sin[\omega_o t + \varphi_i(t)] \tag{6.6}$$

$$u_o(t) = U_{2m}\sin[\omega_o t + \varphi_o(t)] \tag{6.7}$$

where

$$\varphi_i(t) = (\omega_i - \omega_o)t + \theta_i(t) \tag{6.8}$$

$$\varphi_o(t) = \theta_o(t) \tag{6.9}$$

After the ideal multiplication of u_i and u_o, we get

$$u_i(t) \cdot u_o(t) \cdot A_m = \frac{1}{2} A_m U_{1m} U_{2m} \left\{ \sin[2\omega_o t + \varphi_i(t) + \varphi_o(t)] + \sin[\varphi_i(t) - \varphi_o(t)] \right\} \tag{6.10}$$

There are two terms in this product, one being 2ω frequency component. If the multiplier is not strictly linear there will be more high-frequency components of 4ω, 6ω and so on. The high-frequency components being filtered by the following LF, the effective output $u_d(t)$ contains only the phase difference component

$$u_d(t) = K_{PD} \cdot \sin\varphi(t) \tag{6.11}$$

where

$$K_d = \frac{1}{2} A_m U_{1m} U_{2m} \tag{6.12}$$

$$\varphi(t) = \varphi_i(t) - \varphi_o(t) \tag{6.13}$$

u_d varies with $\varphi(t)$ periodically in a sinusoidal way, which explains why it is called a sine wave PD.

When phase difference satisfies $|\varphi(t)| = |\varphi_i(t) - \varphi_o(t)| \leq \frac{\pi}{6}$, sin function can be approximated as linear according to Taylor's expansion. As a result, the expression of u_d could be further simplified as

$$u_d(t) = K_{PD} \cdot \varphi(t) \tag{6.14}$$

The PD is approximated to have a linear response with the phase difference between $u_i(t)$ and $u_o(t)$ within a certain limit. The linearity of PD is demonstrated through the earlier deductions.

A linear system can be analyzed with the standard techniques applied to linear electronic circuit-complex numbers, Fourier and Laplace transforms and can be described by its transfer function $H(s)$. $H(s)$ relates the input and output signals of the system; in conventional electrical networks the input and the output are represented by the voltage $u_i(t)$ and $u_o(t)$ respectively. So $H(s)$ is given by

$$H(s) = \frac{U_o(s)}{U_i(s)} \tag{6.15}$$

where $U_i(s)$ and $U_o(s)$ are the Laplace transforms of $u_i(t)$ and $u_o(t)$ and s is the Laplace operator with the dimension of frequency. We perform Laplace transformation to the time domain eq. (6.14) and get the frequency domain expression or transfer function as

$$u_d(s) = K_{PD} \cdot \varphi(s) \tag{6.16}$$

6.3.2 Loop filter model

As we have discussed previously, the output signal u_d of the PD includes a number of frequency terms; in the locked state of a PLL, first of these is a "DC" component and is roughly proportional to the phase error theta as shown in eq. (6.14); the remaining terms are "AC" components having frequencies of 2ω, 4ω, ...

These higher frequencies are unwanted signals and they are filtered out by the LF. Because the LF must pass the lower frequencies and suppress the higher, it must be a low-pass filter. In most PLL designs a first-order low-pass filter is used. Figure 6.4 lists the versions which are frequently encountered.

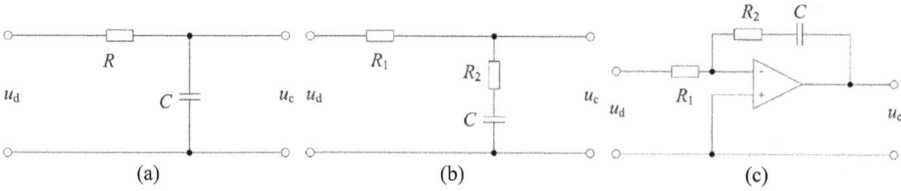

Fig. 6.4: Three common types of loop filter, (a) first-order RC filter, (b) RC integral filter, (c) active filter.

(1) RC integral filter
Figure 6.4(a) is an RC integral filter. Its transfer function is given by

$$H(j\omega) = \frac{u_c(j\omega)}{u_d(j\omega)} = \frac{\frac{1}{j\omega C}}{R + \frac{1}{j\omega C}} = \frac{\frac{1}{RC}}{j\omega + \frac{1}{RC}} \tag{6.17}$$

We convert this expression in the Laplace transform by substituting $j\omega$ by s and get

$$H(s) = \frac{\frac{1}{RC}}{s + \frac{1}{RC}} = \frac{\frac{1}{\tau}}{s + \frac{1}{\tau}} = \frac{1}{s\tau + 1} \tag{6.18}$$

where $\tau = RC$ is a time constant of filter.

(2) Passive lead-lag filter
Figure 6.4(b) is a passive lead-lag filter also referred to as the passive proportional integral filter. A large capacitor is added to break the newly introduced DC path through the R_2 resistor. It has one pole and one zero. Its transfer function is given by

$$H(s) = \frac{u_c(s)}{u_d(s)} = \frac{R_2 + \frac{1}{sC}}{R_1 + R_2 + \frac{1}{sC}} = \frac{s\tau_2 + 1}{s(\tau_1 + \tau_2) + 1} \tag{6.19}$$

(3) Active proportional integral (PI) filter
Figure 6.4(c) is an active PI filter, which is also a lead-lag filter. The term PI is taken from control theory. The transfer function of the active PI filter is given by

$$H(s) = \frac{u_c(s)}{u_d(s)} = \frac{R_2 + \frac{1}{sC}}{R_1} = \frac{s\tau_2 + 1}{s\tau_1} \qquad (6.20)$$

where $\tau_1 = R_1 C$ and $\tau_2 = R_2 C$. The PI filter has a pole at $s = 0$ and therefore behaves like an integrator. It has at least theoretically infinite gain at zero frequency.

6.3.3 VCO

In Fig. 6.1, the frequency of the output signal of VCO $u_o(t)$ is adjusted by the control voltage u_c to force f_o approach to the reference frequency f_i until they are identical with a constant phase difference, hence achieve the lock status of the PLL. It is desirable to have a linear relationship between control voltage and the output frequency. The block diagram is given below in Fig. 6.5.

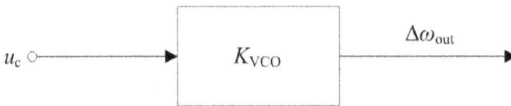

Fig. 6.5: VCO block diagram.

VCO is essentially a voltage frequency converter. A typical device serving as a voltage frequency converter is a varactor. Control voltage u_c causes junction capacity to change together with it, which varies the frequency $w(t)$ of the oscillator. $w(t)$ has a linear response with $u_c(t)$ over a limited range.

$$w(t) = w_0 + K_{VCO} \cdot u_c(t) \qquad (6.21)$$

where w_0 is the center frequency of the VCO and K_{VCO} is called VCO gain with the unit of rad/(V $\cdot$ s). By doing an integral at both side of this equation we get the instantaneous phase as

$$\varphi_{o1}(t) = \int_0^t w(t)dt = w_0 t + \int_0^t K_{VCO} \cdot u_c(t)dt \qquad (6.22)$$

The instantaneous phase of VCO caused by the control voltage u_c is given by

$$\varphi_o(t) = \varphi_{o1}(t) - w_0 t = \int_0^t K_{VCO} \cdot u_c(t)dt \qquad (6.23)$$

After Laplace transformation we get

$$\varphi_o(s) = K_{VCO} \cdot \frac{1}{s} u_c(s) \tag{6.24}$$

The transfer function of VCO is therefore obtained:

$$H(s) = \frac{\varphi_o(s)}{u_c(s)} = K_{VCO} \frac{1}{s} \tag{6.25}$$

6.3.4 PLL model

From the mathematical model analysis of each basic component of PLL, we substitute each transfer function into the overall expression and obtain the linear model of the locked PLL as follows:

$$\varphi_o(s) = [\varphi_i(s) - \varphi_o(s)]K_{PD} \cdot H(s) \cdot K_{VCO} \frac{1}{s} \tag{6.26}$$

$$F(s) = \frac{\varphi_o(s)}{\varphi_i(s)} = \frac{K_{PD} \cdot K_{VCO} \cdot H(s)}{s + K_{PD} \cdot K_{VCO} \cdot H(s)} \tag{6.27}$$

$\varphi_i(s) - \varphi_o(s) \equiv \varphi(s)$ represents the phase error of PLL in the locked state.

6.4 Application of the PLL

At early stages PLLs were used in synchronization systems of TVs to improve the coherence of images. Since 1950s PLLs have been significantly involved in space telecommunication devices such as satellite, space ships for weak signals tracking and reception. The narrow band receivers made of PLLs are excellent in extracting small signals out of complex background noises of the universe. As modern electronics have been developed extensively nowadays, PLLs have become widely used in many more applications such as oscillators, modulators and demodulators, synthesizers, multiplier in microprocessors, decoders. When applied as synthesizers, PLLs are capable of operations such as adding, subtraction, multiplication and division, as well as the function of frequency stabilizer. Furthermore, PLLs are excellent filters with extremely narrow bandwidth and tunable center frequency. Such properties outweigh what conventional devices can do and make PLL a highly desirable choice in modern analogue and digital communication systems. We will expand a few applications here, for example.

6.4.1 Frequency synthesizers

One of the most significant applications of a PLL is the frequency synthesizer. The frequency range of 87.5–108.0 MHz is often assigned for FM broadcasting in many countries. For example, radio stations are anticipated to be hosted within this certain frequency range. It means that each of the stations takes a frequency point and is separated by an increment of 200 kHz from 87.9 MHz to 107.9 MHz. This requires local carrier frequencies. Considering that the crystal is quite large in size, having crystals integrated in one radio would be a very impractical design.

Instead, only one crystal reference is sufficient to provide a stable input reference to PLL that includes a "programmable divider" in its feedback path (Fig. 6.6). The reference frequency is 10 MHz with a crystal of the frequency synthesizer, and the crystal frequency is divided by 50, that is, 200 kHz reference is perceived by the PD. A VCO with frequency range 87.9–107.9 MHz is integrated here to provide the output signal. The output is also taken back through the feedback loop into the programmable divider (N = 439.5–539.5) trying to generate a 200 kHz and also sent to the PD. The outcome and the 200 kHz reference compared by the PD and VCO frequency vary accordingly until these two 200 kHz signals match exactly. Consequently, the output frequency can be precisely tuned and achieved under this mechanism.

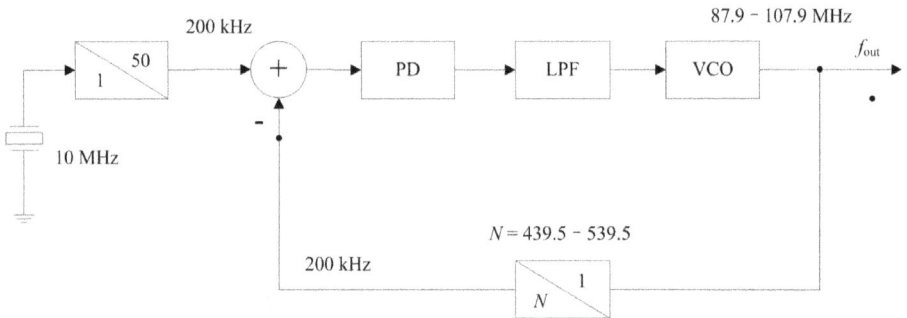

Fig. 6.6: PLL used in FM broadcasting systems.

6.4.2 Local oscillator

PLL circuit can be applied as FM demodulators without fundamental circuit structure adjustments. When the input signal is not modulated which only contains high-frequency carrier wave, the control voltage of VCO is considered at the middle position. When a modulated FM signal is considered as the input, the frequency of the input varies around the carrier frequency all the time as one can imagine. VCO tends to follow this frequency variation and generates an output with the same frequency pattern as the input. The control voltage must change accordingly around the middle position which actually reflects the instantaneous frequency at each moment. The recorded control voltage versus time serves as the demodulated FM signal in principle.

6.4.3 Clock generation

Many electronic systems include processors of various sorts that operate at hundreds of megahertz. Typically, the clocks supplied to these processors come from clock generator PLLs, which multiply a lower frequency reference clock (usually 50 or 100 MHz) up to the operating frequency of the processor. The multiplication factor can be quite large in cases where the operating frequency is multiple gigahertz and the reference crystal is just tens or hundreds of megahertz.

6.5 Summary

In this chapter, we discussed PLL circuit including its basic components, linear model and applications. We expect our readers to have a comprehensive and conceptual understanding about the operational principle. Although linear models are valid only over a very limited range and our content is far from being complete, it should still serve as a beginning of exploration of PLL and help the readers to take the first step into this vast topic. Because the details of PLL circuits are far beyond the scope of this book, we encourage the readers to learn more from specialized PLL references.

Problems

6.1 We gave an example of PLL applied in FM broadcasting systems in Section 6.4.1 and Fig. 6.6. Suppose the FM broadcasting requires the frequency spacing of 300 kHz from 87.9 MHz to 107.7 MHz,
   (1) How should we set the division ratio N of the programmable divider?
   (2) Which crystal oscillator is more favorable, 10 MHz or 15 MHz?

6.2 Please list the commonly used filters in PLLs as well as their transfer functions.

6.3 Explain the conditions and properties for a PLL to be inlock status.

6.4 Why do we use the instantaneous phase of a VCO as its output? Explain the integration function of a VCO in a PLL application.

6.5 Draw the basic block diagram of a PLL circuit and answer:
(1) What is the relationship between output frequency ω_0 and input frequency ω_i when PLL is inlock?
(2) Which physical parameter is compared in a PD?

6.6 Draw the block diagram of a PLL FM demodulator and briefly explain its working principle.

References

[1]   Shen WC, Li X, Chen TM. Communication circuits. Xi'an: Xidian University Press, 2011.
      (only available in Chinese).
[2]   Yang CG. High frequency electronic circuits. Beijing: Higher Education Press, 2006.
      (only available in Chinese).
[3]   Yu HZ. Communication electronic circuits. Beijing: Tsinghua University Press, 2005.
      (only available in Chinese).
[4]   Wang WD. High frequency electronic circuits. Beijing: Publishing House of Electronics
      Industry, 2004.
[5]   Colantonio P, Giannini F, Limiti E. High efficiency RF and microwave solid state power
      amplifiers, 2009. Chichester: John Wiley & Sons, Inc.
[6]   Cripps SC. RF power amplifiers for wireless communications. 2nd ed. Norwood: Artech House,
      2006.
[7]   Niu GF, Cressler JD, Gogineni U, et al. A new common-emitter hybrid-π small-signal
      equivalent circuit for bipolar transistors with significant neutral base recombination.
      IEEE Trans. Electron Devices, 1999, 46(6): 1166–1173.
[8]   Phang CH, Yeow YT, Barham RA, et al. Measurement of hybrid-π equivalent circuit parameters
      of bipolar junction transistors in undergraduate laboratories. IEEE Trans. Education, 1997,
      40(3): 213–218.
[9]   Gu BL. Communication electronic circuits. 2nd ed. Beijing: Publishing House of Electronics
      Industry, 2007. (only available in Chinese).
[10] Zhang YF, Feng JH. High frequency electronic circuits. Harbin: Harbin Institute of Technology
      Press, 2002.
[11] Feng J, Xie JK. Electronic circuits (nonlinear). 5th ed. Beijing: Higher Education Press, 2010.
      (only available in Chinese).
[12] Wang SN, Cheng DH. Electronic circuits instruction book. 4th ed. Beijing: Higher Education
      Press, 2003.
[13] Gao JX. High frequency electronic circuits. Beijing: Publishing House of Electronics Industry,
      2003. (only available in Chinese).
[14] Grebennikov A, Sokalno. Switchmode RF power amplifiers. New York: Newnes, 2007.
[15] Kazimierczuk MK. RF power amplifiers. Hoboken: John Wiley & Sons, Inc, 2006.
[16] Pederson DO, Mayaram K. Analog integrated circuits for communication. 2nd ed. Berlin:
      Springer, 2011.
[17] Robert S. Wireless communication electronics. Berlin: Springer, 2012.
[18] Zeng XW. Communication electronic circuits. Beijing: Science Press, 2006. (only available
      in Chinese).
[19] Zhang SW. High frequency electronic circuits. 5th ed. Beijing: Higher Education Press, 2009.
      (only available in Chinese)
[20] Zhang YF. High frequency electronic circuits. Harbin: Harbin Institute of Technology Press,
      1996. (only available in Chinese).
[21] MC3362 Datasheet PDF (available online).

https://doi.org/10.1515/9783110593822-007

Index

https://doi.org/10.1515/9783110593822-008

www.ingramcontent.com/pod-product-compliance
Lightning Source LLC
Chambersburg PA
CBHW061359210326
41598CB00035B/6041